Advance Praise for

# *Chasing the Sun*

You, too, can help save the world. Read this book!

— Thom Hartmann, author of *The Last Hours of Ancient Sunlight*

Neville Williams did not invent solar cells, but he has tirelessly promoted them. If two billion people in developing countries one day get electricity, it will be because of the groundwork he has laid. Chasing the Sun engagingly chronicles the author's enduring effort to build a solar world.

— Lester R. Brown, president, Earth Policy Institute

*Chasing the Sun* is a fascinating account of the author's personal odyssey to promote solar energy in the developing world. Reading this book, I was reminded of what my good friend, the late Buckminster Fuller once said — "There is no shortage of energy on this planet; but there is a serious shortage of intelligence!". Neville Williams shows how intelligent, local solutions can be found to meet energy needs of communities across Asia and Africa. This should be an eye-opener for donors and policy-makers alike.

— Sir Arthur C. Clarke, inventor of the communications satellite

*Chasing the Sun* is a maddeningly frustrating and extremely inspiring diary of the 15-year effort by Neville Williams and his colleagues to provide solar energy to rural areas of the developing world. The book chronicles the mindless obstructionism of the development establishment, misguided competition from a multinational oil giant and the reflexive skepticism of conventional investment bankers. But the saving grace of this meticulously detailed odyssey is the thrilling enthusiasm of poor farmers and their families in south Asia whose lives have been forever transformed by the gift of low-cost and unlimited energy.

— Ross Gelbspan, author, *The Heat Is On* and *Boiling Point*

# CHASING THE SUN

Philip Jones
Griffiths

Philip Jones Griffiths

# CHASING THE SUN

## SOLAR ADVENTURES AROUND THE WORLD

NEVILLE WILLIAMS

NEW SOCIETY PUBLISHERS

Cataloging in Publication Data:
A catalog record for this publication is available from the National Library of
Canada.

All interior photographs courtesy of Solar Electric Light Fund unless otherwise
credited.
Cover design by Diane McIntosh. Cover photo Joseph Sohm.

Printed in Canada. First printing September 2005.

Paperback ISBN: 0-86571-537-8

Inquiries regarding requests to reprint all or part of *Chasing the Sun* should be
addressed to New Society Publishers at the address below.

To order directly from the publishers, please call toll-free (North America)
1-800-567-6772, or order online at www.newsociety.com

Any other inquiries can be directed by mail to:

New Society Publishers
P.O. Box 189, Gabriola Island, BC V0R 1X0, Canada
1-800-567-6772

New Society Publishers' mission is to publish books that contribute in fundamen-
tal ways to building an ecologically sustainable and just society, and to do so with
the least possible impact on the environment, in a manner that models this vision.
We are committed to doing this not just through education, but through action. We
are acting on our commitment to the world's remaining ancient forests by phasing
out our paper supply from ancient forests worldwide. This book is one step toward
ending global deforestation and climate change. It is printed on acid-free paper that
is **100% old growth forest-free** (100% post-consumer recycled), processed chlorine
free, and printed with vegetable-based, low-VOC inks. For further information, or
to browse our full list of books and purchase securely, visit our website at:
www.newsociety.com

NEW SOCIETY PUBLISHERS                                            www.newsociety.com

*To Patricia,*
*without whom none of this*
*would have been possible.*

# Contents

The most important thing in the world is to have the love of
God in your heart. The next most important thing is to
have electricity in your house.

— Tennessee Farmer, 1940,
(inscribed over the entrance to the electricity exhibit at the
Henry Ford Museum, Dearborn, Michigan)

I'd put my money on solar energy. What a source of power!
I hope we don't have to wait until oil and coal run out before we
tackle that. I wish I had more years left.

— Thomas Alva Edison

The electric light ended the regime of night and day, of indoors
and out-of-doors. In a word, the message of the electric light is
total change. It is pure information without any content to
restrict its transforming and informing power.

— Marshall McLuhan

Solar power provides a marvelous alternative source of energy,
particularly in remote places. It has been proven in the harshest
of environments both terrestrial and extraterrestrial. With solar
power, the village in the jungle can leapfrog into the modern age.

— Sir Arthur C. Clarke

The heart of the matter, as I see it, is the stark fact that world
poverty is primarily a problem of two million villages, and thus
a problem of two thousand million villagers. The solution cannot
be found in the cities of the poor countries. Unless the hinter-
land can be made tolerable, the problem of world poverty is
intolerable, and inevitably will get worse.

— E.F. Schumacher

## Acknowledgments

Many people were involved with this solar odyssey, many helped me, many taught me or inspired me, and many became lifelong friends. I can't name all to whom a debt of gratitude is owed, but I'll try. First, let me express my deepest appreciation to Audrey McClellan, an extremely talented and diligent editor who was assigned to me by New Society Publishers, and who tirelessly went the extra mile to improve the manuscript. I also wish to thank Ingrid Witvoet, the dedicated managing editor who brought the project to completion, and of course Chris Plant and Judith Plant who enthusiastically and fearlessly publish books to save our planet.

I am especially indebted to: Patti Forkan, Bob Freling, Oliver Davidson, Richard Hansen, Walt Ratterman, Tom Tatum, Paul Maycock, Fran Spivey-Weber, Deborah McGaughlin, Henrietta Fiennes, Gerry Gallo, Joel Weingarten, Johnny Weiss, Jeffrey Genzer, Jeremy Rifkin, Jon Naar, Larry LaFranchi, John Corsi, David Grayling, Mike Eckhart, Marlene Brown, Isabella Marks, Maureen McIntyre, Joanne Berkenkamp, Scott Sklar, Ken Jacobson, Mary Fellows, Christine Rubendienst and Marc Koplik.

I would also like to express my gratitude to the following people around the world who made this all happen:

In India: Harish Hande, Kamal Kapadia, M.R. Pai, Thomas Pulenkav, Asavari, Suresh Salvagi, Shridhar Ravapati, Umesh Rai, Hemalata Rao, K.M. Udupa, A.K. Vora, Nitya Mukherjee, the late Vasanthi Pai, Diljeet Titus, Akanksha Chaurey, Dr. V. Bakthavatsalam, B.R. Praabhakara and D.T. Barki.

In Sri Lanka: Priyantha Wijesooriya, Dr. A.T. Ariyaratne, Lallith Gunaratne, Nissanka Weerasekera, Chandima Karunnatilaka, Lal Fernando, Fermando Perera, Susantha Pinto, Sarath De Silva, Anusha Wijeyesekera, Athula Silva, Sonali Fernando, Paul Ratnayeke, Pradip Jayewardene, Sir Arthur C. Clarke, Chandral Chandrasena, Veren Perera, Jayantha Nagendran, Salyia Ranasinghe, Willie Blake.

In Vietnam: Shawn Long, Tran Thanh Danh, Canh Tran, Pete Peterson, Pham Ham Sam, Mme Phuong, Andrew Steer, Dr. Anil Malhotra, Doan Duc Luu, Mme My Hoa, Nguen Tran The, Trinh Quang Dung, Phan Ngoc Chau.

In China: Wang Anhua, Wang Yu, Debra Lew, Charlie Benoit, Mao Yinqiu, Zhang Qiang, Albert Chan, Susan McDade, William Wallace and Scott Vaupan.

In Nepal: Tej Gauchan, Yug Tamrakar, Col. Chhatra Gurung, Adam Friedensohn, Jagan Nath Shrestha, Heshey Phunjok, Dak & Tek Bahadur Gurung, Dinesh Shah.

In Africa: Mark Hankins, Charm Muchenje, Andrew Leitch, Gibson Mandishona, Jurie & Marius Wilemse, Will Cawood, the late Harry Burris, Conrad Roedern, Herman Bos, Robin Moser, John Ssemanda, Izak Kotze.

And in the other corners of Asia and the Pacific: Hermann Oberli, Ford Thai, Peter Banwell, Herb Wade, Rob de Lange, James Finucane, Peter McKenzie, Dr. N. Enebish, Sirikul Prasitpianchi, Maurice Adema, Dipal Barua, Khalid Shams.

In Europe, I would like to acknowledge Wolfgang Palz, Hans Schut, Elisabeth Stern, Georg Fankhauser, Erich Stoeckli, Lars Zoellner, Adolph Kracht, Stephan Schmidheiny, Rolf Gerling, Jeremy Leggett, Bernard McNellis, Tony Derek, Petra Schweizer-Reis, Hermann Scheer, Judith Lipp, the late Sir Kenneth Kleinwort, Lady Kleinwort, Gernot Oswald, Luigi Cuozzo, Rene Magermans, Daniele Guidi, Angelika Baumann, Surekha Aggarwal, Heinz-Wolfgang Bohnke, Edward Goldsmith, Peter Adelman, Mario Posnansky, Philip Wolfe, Peter Ahm, and the late Professor Bob Hill.

At the World Bank, the development agencies, the US Gov't, and at all the NGOs, research institutes, and universities a word of appreciation to: Anil Cabraal, Alan Miller, Louis Boorstin, Richard Spencer, Loretta Schaeffer, Sandra Dolinar, Brooks Brown, Michael Crossetti, Dr. Suresh Hurry, Ernesto Terrado, Jon Exel, Mathew Mendis, Robert van der Plas Jr., Sander Tideman, Bob Schulte, Paul Hassing, Paul van Aalst, Susan Bogach, Maurice Biron, Douglas Salloum, John Thornton, Geoff Stapleton, Jack Stone, David Renne, Lawrence Flowers, Mitchell Strauss, Robin Broadfield, the late Mark Fitzgerald, Eric Martinot, Roger Taylor, Peter de Groot, Michael Totten, Allen Hoffman, Frank Tugwell, Jim Rannels, Richard King, Killian Reiche, Jefferson Seabright, Jerome Weingart, Lester Brown, Michael Philips, Judy Siegal, James Rannels, Larry Kazmerski, Charlie Gay, Eric Usher, Griffin Thompson, Robert "Bud" Annan, Gustav Grob, Dana Younger, Joan Martin-Brown, Stephan Hirsch, Omi Walden, Dr. Peter Bourne, Dr. Fred Morse, Mark Trexler, Daniel Kammen, Mac Cosgrove-Davies.

And the philanthropists who shared the vision, a hearty thank you to: Michael Northrop, Peter Riggs, Robert Crane, Ed Begley Jr., Dr. Al Binger, Wade Greene, Christine Eibs-Singer, Phil Larocco, the late Bob Wallace, David Zucker, Ed Begley, Jon Davison, Jack Vanderryn, and the late Steve Allen.

And a special acknowledgment to all those working for a better world whom I had the privilege of knowing during this journey: Peter Lowenthal, Jorge Huacuz, Chris Flavin, Eric Daniels, Chris Pope, Lisa Frantzis, Ross Gelbspan, Stuart Hart, Craig van Note, John Perlin, Judi & Lou Friedman, Dr. Peter Varadi, Nancy Bacon, Richard Bleiden, John Ewen, Stephan Molivadas, Pete Myers, David Carlson, Mike Bergey, Art Lilly, Mark Farber, Mark Kapner, Les Gordon, Roger Little, Sanford Ovshinsky, Mike Niklas, Carter Hertzberg, Mark Cherniack, Chaz Feinstein, Hazel Henderson, Tim Ball, Denis Hayes, Beth Richards, Larry Rockefeller, the late Harvey Forrest, S. David Freeman, Mark Roberts, Angelina Galiteva, Jorgdieter Anhalt, Les Poole, Chet Farris, Ken Olson, John Naisbitt, Len Jornlin, Steven Kauffman, Carl Weinberg, Mike Allen, Windy Dankoff, Brad Rose, Dr. Chris Sherring, Len Jornlin, Jay Hakes, John Rogers and Peter Yarrow.

# Introduction

This is a book for doers and dreamers. It is about real people doing real things. It is not a science text, a policy treatise, nor an academic analysis of energy and technology issues. It is not a technical handbook. Those have been done, aplenty.

It is a practical "how-to" book: how to change the world, one solar photovoltaic-powered house at a time. Ugh. What an un-elegant word for such an elegant technology! Solar photovoltaics (PV) has been handicapped by this name from the beginning. I prefer "solarelectric" to describe solar power generated by photovoltaic technology that converts sunlight directly into electricity. Or just solar power. I once heard a government official at the dedication of the University of Maryland's solar-powered race car call it "photogalactic." Maybe it's more galactic than voltaic, since the technology does seem out of this world.

Solar energy is generally a boring subject. But getting and using free energy from the sun is exciting and, if humanity is to survive, ultimately necessary. This book is about using solar energy in parts of the world where there are no alternatives other than kerosene and candles. In America, people don't pay much attention to solar power because we have other sources of energy, but when it is your only source of electricity, it becomes pretty interesting — at least to the user. It can also become a business. Or as Shell, the world's second-largest oil company, says, "It may become our biggest business yet." Unbeknownst to most Americans (in spite of the multimillion-dollar advertising campaign launched by the oil giant in 2001), Shell is one of the

1

world's largest manufacturers of solar photovoltaics and sellers of solar power.

Shell knows something most people don't: the world is running out of oil, perhaps faster than we think. And long before we squeeze the wells dry, the increasingly reduced supply will drive prices through the roof, destroying economies, nations, and communities. Thus, energy is the most important issue of our time. What comes after oil will determine how we live on this planet.

So this book is really about energy and how people around the world are using inexhaustible energy sources like sunlight. Ironically, people who never benefited from the age of oil, who never had electricity, are the solar power pioneers who today are using the technology we will all be using in the future. *They* are building the much-touted "solar economy." This book is about them. It is also about the people of the Solar Electric Light Fund, a nonprofit promoter of solar in 11 developing countries, and about the people involved in its spin-off, the Solar Electric Light Company, a commercial venture that, to date, has sold and installed over 50,000 solarelectric systems in developing countries, chiefly India, Sri Lanka, and Vietnam.

Indira Gandhi once said, "There are two kinds of people, those who do the work and those who take the credit. Try to be in the first group; there is less competition there." This book is about the first group, which has received very little recognition for its pioneering efforts around the world, while government policy advisors, academic authors, development economists, and environmental activists often take the credit.

It is also a book of hope, another commodity that seems in increasingly short supply these days as we are faced with "resource wars" and the "clash of civilizations." The lethal mix of oil, Islam, and Israel renders the world a more unstable and frightening place than it has ever been. The hope is found in the delight expressed in a child's eyes as she flips a light switch in her family's wattle-and-daub house for the first time and watches an electric light come on. The hope is represented by the fact that this family was able to purchase their solarelectric installation on credit, and for the first time ever their world did not go dark at 6 PM, as it does year-round in the equatorial latitudes. The hope is in their empowerment.

The hope is that there are clean energy solutions that could work for the whole of humanity. While the West fights its wars over oil, humble farmers in so-called developing countries are already putting this 21$^{st}$-century solar solution to work. However, few Americans know about it because this energy revolution is taking place so far away, and until now, not much has been published about it. There is hope in knowing that it will take place here too, in young people's lifetimes. At 62, I probably won't see it, but I can assure you that it is coming.

If you believe, as I do, that energy is the world's Topic A, then you will enjoy reading about the human side of promoting, delivering, selling, and using solar electricity. You don't need to know, or care, anything about how this technology works, any more than you need to understand the principles of a cathode-ray tube or a liquid crystal display to turn on and watch a television. You don't need to be able to tell photons from electrons to appreciate the attraction of solar energy. It does not take technical knowledge to be a sun worshipper.

This book tells the story of how a small group of doers and dreamers struggled to turn their "unrealistic" vision into the reality of a quarter of a million people getting their household energy from the sun. And it asks the question: If the poorest people in the world can have a "solar solution," why can't people in North America?

After oil, we will still need energy, and this is it: solar electricity, solar energy, solar power. The sun is ready. Are you?

# CHAPTER I

# Solar Revolution?

It seems like a dream now, the halcyon days of the 1950s, when I was growing up in a time of innocence in a world full of promise, without cynicism or irony. The bright future lay before me, around the corner, capturing my imagination. There was no fear in the air, no shortages of anything, no worries, really. To a high-school student in a small Ohio town, whose father's metallurgical engineering job paid for a large white house on three acres, a new '57 Chevy, and a second car for his mother who didn't need to work, it was the best of times.

I was born in 1943, at the beginning of the baby boom, when hundreds of thousands of soldiers left their wives (soon-to-be mothers) behind as they shipped off to war. (No, the baby boom didn't start when the men came back from war in 1946, but when millions left for the European and Pacific theaters in 1943.) The cultural icons of the boomer generation — Bob Dylan ('42), the Rolling Stones (Mick Jagger, '43), and the Beatles (Paul McCartney, '42) — defined the zeitgeist as one of fun fun fun, wild rebellion, and indignant protest. None of us was to have a normal life.

It wasn't until I was in high school that I learned we'd won World War II by dropping two nuclear bombs on Japanese industrial cities, obliterating them and some 200,000 of their inhabitants. By this time, Russia had the bomb as well. I still remember bicycling six miles to town to buy an issue of *Mad* magazine that showed Planet Earth with big chunks blown out of it as Russia and the United States lobbed ever-larger intercontinental ballistic missiles back and forth at each other. Ha ha.

5

However, the threat of nuclear missiles was offset by President Dwight D. Eisenhower's Atoms for Peace program, which pumped billions of government dollars into building the first nuclear power plants. Soon, it was said, we would have power "too cheap to meter." You just needed a little uranium and some plutonium 235 and you'd have all the power you'd ever need, forever and ever.

This seemed too good to be true, even to a 10th grader, so I wrote to the Atoms for Peace program in Washington, and soon a huge package of technical materials arrived in our mailbox. I studied them carefully, my father helped me with the technical terms and descriptions, and soon I understood how nuclear energy was produced. Amazing! Neat!

What could be better than this?

I decided to produce a detailed, comprehensive, schematic drawing of exactly how a nuclear power plant worked for my 10th-grade science project. I mounted the drawing on a big wooden board, carted it off to school, and explained to the science teacher all about critical mass and controlled reaction and heat transfer to make steam for generators, ultimately producing *electricity*.

It was many years before I learned that it would be better to use photons from the sun to convert electrons to electrical energy. It would be every bit as clean as nuclear power, without the risk of contaminating our cities for 25,000 years if something went wrong (and something will always go wrong, some day), and it would eventually prove to be cheaper than nuclear energy, which became so costly that it nearly bankrupted several US utilities that invested too heavily in this energy source.

The sun was the "natural source for electricity," to borrow BP Solar's marketing motto at the turn of the 21st century.

Despite this early interest in electricity production, I did not get a job in the power industry, nor did I study science, engineering, technology, or business. (At the University of Colorado in the 1960s we regarded business students as losers, one step below those taking basket weaving or phys ed.) None of my father's astonishing abilities in mechanics, science, and engineering rubbed off on me. I decided to become a writer and a journalist; studied history, English, and "poli sci," and set out after college to cover the story of my generation: Vietnam. I decided to "learn by doing." I learned to write about war by going out to see it and watching my contemporaries fight it. (I was drafted, too, but that's not a subject for this book.)

The Vietnam War alienated a generation, and many of us "dropped out" of mainstream society. I chose the freelance writer's life in a Colorado mountain town, a place to recover from the '60s and try to become whole again. Meanwhile, the Middle East soon replaced Vietnam as a focus of our national attention.

After the 1973 oil embargo caused by the Middle Eastern members of the Organization of Petroleum Exporting Countries (OPEC), *energy* became dinner table conversation. Soon thereafter, in the wake of the Watergate scandal, Nixon resigning, and the United States hightailing it out of Vietnam at long last, losing our first war, nuclear scientist and nuclear submarine captain (and Georgia governor) Jimmy Carter was elected to preside over a very depressed United States, wearing his cardigan and telling us to turn the thermostats down ... to save energy.

President Carter created a new government department, the United States Department of Energy (DOE), and cabinet post. The DOE included a division dealing exclusively with energy conservation and solar energy, headed up by Carter confidante Omi Walden, the first assistant secretary for Energy Conservation and Solar Energy in the US government. She grew up on a farm in Georgia that did not have electricity, but did have a windmill to generate power for a battery that could run lights to illuminate the farm at night.

Tom Tatum was one of Carter's bright young campaign operatives, a Georgia native and Vanderbilt University law graduate, who earned his political bona fides managing Maynard Jackson's campaign to become the first black mayor of Atlanta. Carter wanted to keep a close watch on his pet project, a new energy policy, so he brought Tom to Washington and offered him a special assignment as the key political liaison under Walden at the DOE's Office of Energy Conservation and Solar Energy (which was soon known as "Conservation and Solar").

Tom met me in the aforementioned Colorado mountain town, where he had recently bought a ski condo. Our mutual friend Sam Brown, a former antiwar activist who was now working for Carter in Washington, had told Tom to look me up. We met at the local saloon, where sex, drugs, and rock 'n' roll were the usual subjects of conversation. Instead, we talked serious politics. Scotch, our drug of choice, lubricated late-night conversations about ... energy! Tom said he needed help in Washington at the new Department of Energy to launch "the solar revolution."

The what?

Licking my existential wounds from the Vietnam War, and tired of skiing one of the world's magnificent ski mountains seven days a week, I was ready for a good fight, and a *revolution* sounded perfect. So when Tom, back in Washington, called to ask me to come to DC immediately to help him promote solar power and energy conservation at the highest levels of the US government, I said sure. I came down from the mountain, spent part of 1979 in Washington at DOE, and returned for the summer of 1980, still the hottest on record. Temperatures hovered at 86 degrees *indoors* because President Carter had ordered all the government thermostats set at 80 to save energy. And you couldn't open the windows.

Despite the heat, no one had ever heard of global warming. That would come later. What we were concerned about, talking late into the night over our Dewar's at the Hawk & Dove on Capitol Hill, was *energy security*. Carter had come to power just after the United States had passed "peak oil," the point at which the country began to extract its oil reserves faster than it could find new oil deposits. The United States was no longer self-sufficient in the oil department, and we began to import more and more crude oil until we were importing more than 50 percent of our daily needs from the very OPEC countries that had caused the long gas lines in 1973 and who were now (1979) raising prices to over $30 a barrel. In July 1979, President Carter addressed the country and said, "Energy will be the immediate test of our ability to unite the nation."

The issue quickly became: How could the United States wean itself off dependence on foreign oil? Especially oil from the unstable Middle East?

Tom and I thought we knew. We read S. David Freeman's *Energy: The New Era*. He was a guru to us. Carter had appointed Freeman to head the Tennessee Valley Authority, where he was still completing 17 nuclear plants, but preaching energy independence and promoting other sources, including solar. He would eventually close most of the nuke plants down. (David and I later crossed paths in a remarkable, if unfortunate, way, as I explain in Chapter 5.) Denis Hayes, the founder of Earth Day in 1970 (with Senator Gaylord Nelson), influenced us as he wrote about the "soft energy path" — meaning "renewable energy" (which was a new phrase for me) — and the brilliant young Amory Lovins was agitating for "energy conservation," making his views known at our office in DOE. Tom directly advised the president about these new policy ideas.

I was well equipped for the job when I moved into my office next to Omi Walden at DOE and began writing her speeches and developing policy briefs. I had a US government-issue manual typewriter, as I never could manage to work an IBM Selectric. Computers? Please. I'd only seen one, and it took up about 700 square feet. And I knew absolutely nothing about energy conservation or solar energy.

Not knowing what I was doing has never stopped me. This should be a lesson of inspiration; you can do anything you want to do as long as you don't know anything about it. Knowing too much stops you in your tracks.

I was interested in this so-called solar revolution, whatever it was supposed to be about. And as a red-blooded, patriotic American, I didn't like the idea of OPEC countries controlling our economy, our future, America's destiny. We had to find "alternative sources" of energy, whatever those might be.

Before I continue, let me say that if this was a book about energy policy, I could stop right here because absolutely nothing has changed in the United States in the past 25 years regarding energy policy. Today we're debating "energy security" exactly as it was debated by us young hotshots decades ago in the bars and cloakrooms of the nation's capital, except that then — with visionaries like Senator Paul Tsongas and Congressman Richard Ottinger — we helped the administration come up with a national energy plan that was far more progressive than anything seen in the Congress since, despite the fact that now almost *60 percent* of our oil comes from overseas. And in the wake of the Gulf War, September 11, and the Iraq War, energy security is far more critical than it was during the Carter era. *Plus ça change, plus c'est la même chose.* Or is it *déjà vu* all over again?

And the environment? No one thought about it then. Today we think about it a lot, but do nothing, while *energy security* has trumped concern over greenhouse gases and global warming. The right of Americans to drive SUVs trumps concern over the loss of young lives in Lebanon (250 Marines in 1981), Kuwait, Yemen, and Iraq (1860 and counting as I write this) as the United States seeks to keep the oil regions stable.

Meanwhile, back at DOE: President Carter came down from *his* mountain —Camp David — like Moses, clutching his national energy plan and proclaiming for all the world to hear, "This democracy which we love is going to make its stand on the battlefield of energy." He said that this would require the "most massive peacetime commitment of funds and resources in the nation's history." The $88 billion package Congress subsequently approved for Carter's energy initiatives was bigger than the Marshall Plan for the reconstruction of Europe after World War II.

On June 20, 1979, Carter gave his "solar energy" speech to Congress, outlining a "national solar strategy" that included the goal of supplying 20 percent of our nation's energy needs from the sun by the year 2000 ("from the sun" included hydro, since the sun makes rain that makes water for hydropower). Carter had announced his "domestic policy review" for solar energy the year before on "Sun Day" in Denver. Tom and I were ecstatic; everything we were working for had the full backing of the president of the United States, or so we thought.

Then we discovered that more powerful people than our little group at Conservation and Solar had the ear of the energy secretary, Jim Schlesinger (who had been defense secretary under presidents Nixon and Ford). These were, you guessed it, the oil companies, mainly Exxon and Occidental. To Secretary Schlesinger's credit, he did say, shortly after taking the job, just after the near-meltdown at the Three Mile Island nuclear facility, that "no nuclear plant will ever get built in the USA again," and he was right about that. Nuclear was dead. But that didn't mean he believed the president's goal of a 20 percent renewable-resource-based society was attainable, or that he'd throw the previously gigantic subsidies for nuclear over to solar and renewables.

That $88 billion was largely going for the "alternative energy" called ... *synfuels* (synthetic fuels). Does anyone remember synfuels? This was oil "mined" from rock by heating it in situ. Exxon, ARCO, and Occidental Petroleum raced to extract oil from the vast oil shale deposits of western Colorado. Exxon was going to strip-mine for oil, and Occidental drilled one of the world's deepest mine shafts, looking to bring it out that way. Did it matter to anyone that extracting oil from shale was more costly than any already proven solar technology, thermal or electric? Not in a country ruled by "big oil."

"The old nuclear crowd and the fossil-fuel guys within the department were really astonished that the president made that kind of commitment to solar," Tom told me later. "The sandbagging started immediately. I tried to get the White House to allow the president's solar speech before Congress to be televised, but they said no, and it went by like a speeding bullet."

On the plus side, the Carter energy plan included substantial funds for Conservation and Solar at DOE, including research money for the new Solar Energy Research Institute (SERI) at Golden, Colorado, to be headed by Denis Hayes. We still had plenty of work to do on the solar front. We never believed in the "synfuel fix," and I wrote numerous articles exposing the program. Ironically, it was President Ronald Reagan's people who pulled the plug on it, recognizing it as a huge corporate giveaway. Not all the oil companies got their hands dirty with oil shale: Atlantic Richfield (ARCO), under the leadership of Robert O. Anderson, sold its interests to Exxon. Anderson saw it for the scam that it was. His was the first oil company to invest in solar.

Since none of us was sure if President Carter would get a second term, given his standing in the polls in 1979, we decided we had to act quickly if solar energy was going to see the light of day, so to speak. We had the blessing of the White House, even if the aging "old technology" guys in DOE were trying to kill us. I thought of German physicist Max Planck's quote: "New ideas in physics triumph only after the adherents of the old ideas have died."

The most exciting new idea in physics was *photovoltaics*. A terrible name for a wonderful, magical, elegant technology that Bell Labs first developed in 1953. Photovoltaics, or PV, as it came to be known, refers to the production of electricity from sunlight using "solar cells." Up to 36 of these small, round, silicon wafers are soldered together with metal tabs and assembled into a "module," commonly called a solar "panel," which converts sunlight into electrons. The electrons flow as 12-volt direct current from the solar cells. Asking me to explain this would be like asking a television sit-com producer to explain how TV works (see Chapter 4).

Captain Paul Maycock, a physicist formerly with Bell Labs and Texas Instruments, headed up the PV shop at DOE, managing a $900-million applied research and "commercialization" budget under the Carter program. He was called "Boomer," for he spoke in a voice that filled up a room of any size without amplification. Rotund, bearded, and intensely absent-minded, he was busy bringing government support to the fledgling technology through contracts with manufacturers, research labs, and commercial innovators.

Paul told me, "Residential systems are being sold all over the world right now. Solar cells are already economically viable for off-the-grid, isolated homes and villages. By 1986, we expect photovoltaics to be fully economic for residential use in this country, at a cost to consumers of six cents per kilowatt hour in the sunnier parts of the country."

Paul was dreaming, as we'll see, but so were we all.

Denis Hayes at SERI out in Colorado was predicting that we'd achieve DOE's goal of bringing the cost of solar cells down to 70 cents per watt by the mid-1980s. If we had done that, I wouldn't have to write this book. In 2003, PV module prices averaged $3 per watt, but that was a lot better than $90 a watt, which is what modules cost in the late 1970s, and a big drop from the $10 per watt price for the PV modules used in the world's first 100 kW array installed at Natural Bridges, Utah, in 1980.

I was excited. Here was the technology that could save humanity! It was 100 percent American ingenuity. We could make enough solar cells from silicon, the most abundant element on earth after oxygen, to not only replace oil and coal, but to build decentralized, self-sufficient power networks managed by microprocessors that would allow every home to be its own utility. Suburban roofs would be covered with solar panels, and huge "arrays" of panels in the deserts of Arizona and California would produce industrial-strength power for the grid. Cars would all be electric (DOE's electric-car people predicted that 9 million all-electric cars would be on the road by 1990), charged up by the sun.

We could bring about the "demassified" society that Alvin Toffler talked about in *The Third Wave*, make families "energy independent," and foster self-reliance and a "decentralized" energy system that author Jeremy Rifkin and others would advocate 20 years later.

Photovoltaics, coupled with wind energy, small and large hydro plants, biomass, cogeneration, energy conservation and efficiency, solar thermal plants, solar water heaters, and whatever else the busy denizens of DOE's solar office were coming up with, would be the backbone of the future solar economy.

America would be a land of "ecotopians," to borrow a word from Ernest Callenbach's visionary tract, *Ecotopia*.

Free energy from the sun sounded good to me. But we were way ahead of our time, and what we didn't know was that the technology to accomplish all this wasn't quite as ready or as economically viable as we hoped.

Tom asked me to help communicate this dream of a new future to the American people in what would become the largest solar energy communications campaign in American history. Actually, it was the only one.

Tom, six foot three, charismatic, handsome, and irresistible to DC's population of attractive single women, focused on his calling like a Southern preacher. He stood out like a Rhett Butler among the gray hordes of Washington bureaucrats who seldom smiled and never laughed. He had fun, strolling by the sidewalk cafes of Capitol Hill with a different woman on his arm every evening, but by day his dazzling intellect was applied to only one thing: saving the world with solar energy.

Carter, purposefully or accidentally, had given him carte blanche to try. Tom's one-of-a-kind Office of Institutional Liaison and Communications ("I made it up," Tom told me) meant he didn't need to work through DOE's 80-person office of public affairs, which was controlled by ex-military types looking after the *real* concerns of DOE — making nuclear weapons. (The department was also known as "the bomb factory." When Carter formed DOE, Congress dumped the Atomic Energy Commission into it. The commission oversaw the government's breeder reactors, which manufactured fissile material for nuclear warheads, "atoms for peace" having somehow been forgotten along the way.) Tom was ready to take risks to launch a full-blown national promotion of solar energy, using all the budget at his disposal and damn the torpedoes.

We believed it was our job to communicate the new ideas, including how energy conservation could reduce the need for imported oil, what a solar-powered economy might look like, and how we could attain it. There was an air of "national crisis" in Washington; the big worry was that oil could go to

$100 a barrel. Such a price would destroy the world's economy. It had nearly hit $40 in the 1970s before retreating.

The American people had to be told the truth: that truly alternative and energy-saving technologies existed that could lead the transition to a renewable-resources-based society. We had to raise the consciousness of the American people, and we had the platform and resources to do it from our base at DOE. What more did we need?

Hollywood! We'd use entertainment, television, celebrities, movie stars, and the cinema to get the word out, just like the government did to rally America against the Nazis in the early days of WWII.

But first we had to come up with a strategy. I suggested a high-level national retreat somewhere. Tom found the retreat venue: a glorious mountain camp above Boulder, Colorado, a mile and a half above sea level. We didn't want to know the source of the owner's income or how he managed to maintain this plush hideout in the foothills of the Rockies, but when he offered us the Gold Lake Ranch compound for free over beers at the Hotel Boulderado, we took it. The owner, a "Boulder hippie" with business acumen, sought to operate a respectable mountain conference center and he wanted to do his part for the solar revolution. The DOE would be his first big client. He only asked that we pay for the food. Tom okayed the deal, and I was ordered to come up with an invitation list for the Gold Lake Solar Energy Media and Communications Strategy Conference.

We brought together young leaders and professionals from television, film, advertising, banking, politics, local government, architecture, publishing, construction, broadcasting, education, philanthropy, and business. I invited Carl Rogers, an old high-school friend, who had co-founded Vietnam Veterans Against the War, managed the legal defense for Daniel Ellsberg (the man who, in the early 1970s, leaked the "Pentagon Papers," a top-secret White House study on Vietnam decision making), and was currently a media strategist and activist with the No Nukes movement in Los Angeles, which was the hot new cause for the veterans of the anti-Vietnam War movement and the Los Angeles rock music community. Our little DOE workshop was suspect among members of the No Nukes Movement, since it was the DOE that had promoted and subsidized nuclear power and also operated the country's two uranium-processing plants.

I met with the No Nukes organizers at their office on Sunset Boulevard and explained, "No, we're not pro-nuclear. We're from the Office of Conservation and Solar, and we're reporting directly to the president."

Stopping nuclear power was a fine cause, along with reducing the use of imported oil, but what was going to replace these energy sources? Well, the No Nukes people had been using a song by Daryl Hall and John Oates to promote the alternative: "Give me the Warm Power of the Sun." A local musician sang it before the great fireplace at the Gold Lake Long House following our welcome dinner, and we all dreamed of a new America powered by solar energy. (I wouldn't hear the song again until Peter, Paul and Mary sang it from the stage at Earth Day 2000 in Washington, where actor Leonardo DiCaprio and Vice President Al Gore spoke passionately about the environment, standing beneath a 200-foot-long banner proclaiming "Clean Energy Now." Peter, a friend for the past 25 years, told me backstage, just before the trio walked out to the mikes before the huge audience, "This is for you!")

The singer at the Gold Lake Long House was followed by a Ponca Indian poet named Sa-Su-Weh, who recited his poem "Master of the Sun" and then offered around his peace pipe.

What were we smoking? I now ask figuratively. Actually, we were only drinking. No drugs were allowed at the retreat, so we had to be content with the clear Rocky Mountain air, French food, and large quantities of booze. I never knew who picked up the bar tab for what the subsequent DOE report called "an eclectic gathering of professionals."

The goals of the conference, as stated in the Gold Lake Project Report, were "to develop a communications strategy aimed at overcoming the institutional and psychological market barriers to solar energy in this country, and accelerate the transition to a renewable resources society including a 20 percent solar America by the year 2000; to form an advisory pool of non-governmental professional media resources; and to urge the formation of a private-sector national media coalition for the purpose of developing solar markets and prompting the use of the sun's energy."

*What were we drinking?*

Reading the DOE's 16-page summary report a quarter century later is to sadly recall a dream unfulfilled. It's especially melancholy for me, since I wrote it. But Gold Lake *did* launch a national campaign to promote energy alternatives and energy conservation. And the Hollywood communications effort was a direct outcome.

I moved from DC to "duty station Hollywood," where I'd once paid my dues, like so many young writers, trying to get a feature film made. Soon Tom and I were meeting with all the big names in television series production, convincing them to help America solve its energy crisis. Gold Lake had opened a lot of doors. Energy conservation messages were written into dozens of sit-coms, which reached tens of millions of people. I brought in up-and-coming talent like producer Jon Davison ("Airplane," "RoboCop"), and director Ron Howard ("Apollo 13," "A Beautiful Mind"), and Tom fronted a quarter million in DOE money for "Reach for the Sun," a TV film for children about solar energy and energy efficiency, coproduced with KCET-TV, Los Angeles. We worked with Robert Redford to help promote his "Solar Film," a seven-minute masterpiece of solar propaganda offering a new energy future. It ran in thousands of theaters that year as a public interest trailer. Today it is a curious artifact of a long-lost vision.

At Gold Lake we filmed a US government public service announcement featuring Sa-Su-Weh in a tipi teaching his small son about the benefits of the sun's energy. We intercut his solar paean with images of solar hot-water installations on Boulder apartment houses and wind generators in the Midwest. This was followed by dirge music over aerial shots of a supertanker making its way to US shores with its cargo of liquid black gold. Then the words "EXPLORE SOLAR ENERGY, US Department of Energy" appeared on the screen, followed by an 800 number for information.

Several thousand of these "Master of the Sun" PSAs were distributed and played for years at 3 AM on hundreds of local television stations. We also produced a thousand radio PSAs featuring Hall and Oates' "Give Me the Warm Power of the Sun." When the Reagan transition team swept through the DOE in the winter of 1980, they found this subversive material and burned the lot of it. President Reagan preferred the warm power of imported oil (he was right; it was cheaper to get it from Saudi Arabia than from oil shale or renewables). He also had no truck with energy conservation. "America didn't conserve its way to greatness," he was fond of saying.

Actually, it was energy conservation and its counterpart, energy efficiency, that together accounted for an unparalleled energy success story arising from the Carter years. Five years after Jimmy Carter's one-term presidency, the country was using measurably less electricity than when he came into office, despite a healthy growth in the GNP along the way. This was attributable almost solely to government-sponsored energy-saving programs in partnership with industry, business, and consumers. Omi Walden's Office of Conservation and Solar instituted virtually all of the programs that, controversial at the time, became commonplace in the decades after, such as using compact fluorescent lightbulbs; reducing energy waste in appliances, especially refrigerators and air conditioners; making buildings more efficient; implementing industrial cogeneration; and mandating household weatherization. The first solar tax credits proposed by DOE at this time were enacted by Congress, fostering a whole new solar water-heating industry. All this was Carter's big unsung legacy.

Tom's Institutional Liaison and Communications Office led the way with public education, building awareness nationally of the need to "save energy." While we had not made any headway in "selling solar" to the American people — maybe we *were* dreaming on our "solar high" up there at Gold Lake — we were unexpectedly successful at selling energy conservation and efficiency. We launched national low-cost public education and awareness campaigns, brought in the mainstream press for special seminars inside DOE, and worked with the White House to produce the first "energy fair" on the Mall in Washington. Even DOE's public affairs office, under Jim Bishop, a former *Newsweek* reporter, came up with a national campaign with our help: "Energy: We Can't

Afford To Waste It." And the Hollywood effort, launched at Gold Lake, continued to pay off as even soap opera stars, seen by tens of millions of daytime viewers, started talking about energy conservation on and off the screen.

But on the solar front, we lost the war. After Gold Lake, Tom's office proposed to the White House a $50 million paid-advertising campaign for solar energy, but it was denied funding by the appropriations committees. By this time, Tom was getting pushed aside by the oil interests (and the synfuels people) at DOE. They managed to get his independent communications responsibilities transferred to the DOE public affairs office, which buried the solar message for good when a new energy secretary, Charles Duncan Jr., former CEO of Coca-Cola, replaced Bishop with a military guy.

"Duncan brought in 19 special assistants who were trampling all over everybody at this agency," said Tom, "and they spent a great deal of their time trampling over everyone at Conservation and Solar. All of them came from the defense department or industry, with no experience in energy at all."

Tom was incensed that a larger and larger portion of DOE's budget was going for the nuclear warhead program instead of domestic energy needs. "This agency needs to be broken up," he said. Even the *Washington Post* agreed in an editorial at the time. It never was. Reagan said he'd dismantle the DOE when he was elected. He never did. He couldn't. It remains a bloated bureaucracy that even enlightened secretaries like Hazel O'Leary and Bill Richardson could not fix.

At the same time, Congress continued to allocate sizeable funds for solar and conservation. Paul Maycock's shop got $90 million in 1981 to continue research in, and "commercialization" (a term of bureaucratese I learned at DOE) of, photovoltaics. SERI got its operating money, but no money for public education, so few Americans knew that Denis Hayes was running a US government program devoted to solar energy (not conservation or efficiency) that employed nearly 1,000 people out in Colorado. We promoted SERI's solar development activities as best we could.

Thanks to the Middle East stew of oil, Israel, and Islam, which was driving and controlling US foreign policy even before the Cold War ended — and hence determines the outcome of most of our presidential elections — Carter was in trouble. This time it was the Iran hostage crisis, which Reagan said he'd solve, and did. (The Iranians could have elected Carter, if they chose to, by releasing the hostages on his watch.)

Tom, by then getting further sidelined at DOE, saw the writing on the wall. He was disappointed that efforts to get the solar, conservation, and efficiency message out to the American people were being dismantled by DOE officials and that Carter's White House would not intervene.

Tom Tatum resigned from DOE in September 1980, citing "substantive disagreements" with the department's "balanced energy agenda." In his resignation letter he pointed out his opposition to the $88 billion subsidy for "synfuels" technology. "I believe," he wrote, "the resources that have been arrayed to develop the synthetic fuels industry misallocate capital that is badly needed to implement comprehensive energy efficiency programs in the United States and accelerate the use of renewable resources to reach the President's goal of 20% solar by the year 2000." He also based his departure on his opposition to the nuclear weapons programs at DOE and further stated, "I plan to work actively, once I leave government, in the effort to curtail the use of nuclear power in the United States and the world .... There has been no solution to the nuclear waste problem, and perhaps there may never be one."

Tom held a huge farewell party on the roof of the Hotel Washington in September 1980, attended by many of the city's young movers and shakers and dozens of the most beautiful and intelligent women in Washington.

"I'm moving to LA to make movies. They have even *more* beautiful women out there," he told me. He married a lovely film editor, and Tom and I later collaborated on a movie treatment about a conspiracy to stop a technical "breakthrough" in photovoltaics that would have badly hurt the oil companies. Needless to say, the movie didn't get made, although a feature film with a suspiciously similar plot was produced a couple of years later starring Marlon Brando and George C. Scott. It bombed. Tom turned to extreme-sports action films, which he produces to this day.

Paul Maycock, head of the photovoltaics office, left DOE when the Reagan people took away his budget for commercialization of PV, leaving only a small sum for pure research, even though this technology was ready for the marketplace. Had photovoltaics received the same push into the marketplace that nuclear power enjoyed, we'd probably all be living in solarelectric houses today, but Paul was told that, unlike nuclear power and even coal-fired power generation, strongly supported with ten times the subsidy clean-energy technologies received at DOE, solar would have to make its own way in the market.

Paul resigned and started *PV News*, the "insider" photovoltaic newsletter. Exactly ten years later he joined my board of directors at the Solar Electric Light Fund.

Denis Hayes left SERI, which was renamed the National Renewable Energy Laboratory, excising the nasty word "solar."

Our goal of a "20 percent solar-based society by the year 2000" had been a self-inflicted delusion.

I went back to Colorado to escape the Reagan years and became director of marketing for a large ski resort. I didn't think about solar energy for nearly a decade. Later I became a real estate developer of sorts, buying and rehabilitating a historic building and running several businesses from the premises. But all the time I knew I couldn't grow old in a ski town, and I missed having a mission.

Skiing, ski marketing, and real estate development hadn't replaced the heady anti-Vietnam war protests nor the mission-driven effort to solarize America. Magazines I was writing for went out of business — *New Times*, *Rocky Mountain*, *Harper's Weekly* — and I was embarrassed, frankly, to continue contributing to *Penthouse*, which was becoming too explicitly raunchy to show my mother. Although I had written the cover story for the premier issue of *Outside* magazine in 1977, I had not sought a career in outdoor, sports, or adventure writing. *Outside's* first issue contained another feature, about the "eco-commando" environmental organization Greenpeace, founded in 1971 by a small anti-nuke group in Vancouver. It stuck in my mind for the next decade.

In 1987 I landed back in Washington to take a low-paying, challenging job at Greenpeace USA as their national director of media and public relations. Here I met Greenpeace's chairman, David McTaggart, a highly energetic Canadian who had lived and worked and developed real estate in California and Colorado ski resorts before heading off to the South Pacific in a small boat, where he managed single-handedly to end French nuclear testing in the atmosphere. This was a career switch if there ever was one! We had a lot in common. I liked David immensely, and years later I visited his magical house in an ancient olive grove in Umbria, where he asked me how to put solar hot-water and solarelectric systems on his roof. Before we could proceed, however, he was killed in a car crash in Italy, a catastrophic loss to the global environmental movement.

It was at Greenpeace where I first heard the phrase "global warming." I learned a great deal at this amazing organization, which was "campaigning" on just about every environmental issue there was, from nuclear testing (below ground) and nuclear power to toxic wastes, cleaning up rivers and oceans, saving marine mammals, ending the seal slaughter in Canada, and stopping the hunting of whales.

In my new job I was determined to do everything possible to make Greenpeace as famous as it deserved to be. When the French government sank the Greenpeace anti-nuclear campaign ship *Rainbow Warrior* in New Zealand in 1985, the organization's profile soared in Europe. I wanted to do the same thing in the United States.

I called Tom Tatum in LA and brought him back to DC so we could co-produce "Greenpeace's Greatest Hits," a video featuring the organization's first decade of activism. I asked actor John Forsythe ("Dynasty"), who possesses perhaps the greatest voice of the 20th century, to narrate it, and I signed an upcoming "new age" musician, Peter Kater, to write an original score. This was all great fun, and the video was a success, but I was more affected by the negative messages emanating from all the young activists at Greenpeace who saw the world in stark, even hopeless, terms. Greenpeace was about problems, not solutions or hope. And solar energy as a positive partial solution to the energy and greenhouse gas emission problem, and hence global warming, was not on its radar screen.

Five years later Greenpeace International, based in the Netherlands, launched a solar campaign. Geophysicist Dr. Jeremy Leggett of Oxford University advised the group on the need for radical change in energy policy. But in the United States, Greenpeace refused to consider a solar energy campaign until the late 1990s.

BP Solar

Meanwhile, I had begun to wonder, "Whatever happened to solar energy?" Almost nine years had gone by since I left DOE's employ.

I decided to quit Greenpeace and leave behind its vision of gloom and doom. I was looking for hope. I began consulting for Solarex, based in Frederick, Maryland, in a large slab of a building, clearly visible from the highway, that was dramatically covered on

BP Solar's Maryland manufacturing plant, originally Solarex (1973), with 300 kW of grid-tied photovoltaics on its slab roof.

one side by blue polycrystalline solar photovoltaic modules gleaming in the sun. Started in 1983 by Drs. Peter Varadi and Joseph Lindmeyer, who shared a vision of "bringing solar down to earth," Solarex was the largest US PV manufacturer, producing solar modules for terrestrial use as well as for the US space program, which is where PV got its start. The terrestrial market was growing by 30 percent a year.

Whatever happened to solar energy? Well, during the Reagan years the oil companies had bought it, and it was progressing quite nicely, thank you. Solarex was sold to Amoco, which made my friend Peter Varadi, Solarex's co-founder, a very rich man.

Jumping ahead a decade, to the millennium, I sat through an amazingly boring roundtable workshop at the World Resources Institute in Washington, where representatives from numerous nonprofit organizations and their young energy policy wonks endlessly debated whether or not the "energy policy and environmental community" should call for America to derive "10 percent or 20 percent" of its energy from renewables by the year ... *2020!* The group decided that it would only be realistic to unite behind a target of 10 percent renewables by 2020. I found this painful and appalling.

In November 2003, Paul Maycock reported in *PV News* that presidential candidate Howard Dean had called for "a strong renewable energy bill," which unfortunately was not going to happen under the Bush regime. Paul quoted Dean: "The U.S. used to lead the world in this kind of technology, now we are falling behind. We need to make a serious investment in renewable energy. It's important for our security, it's important for our environment, and it's important for our economy. We should be aiming even higher — for a 20 percent standard for 2020."

Paul wrote that "despite pleas from a bipartisan group of 53 senators, the Republican leaders of the House-Senate conference have eliminated a provision from the energy bill to require that our nation generate *ten percent* of its electricity from renewable resources by 2020."

At least Carter got his 20 percent goal included in his energy bill *25 years earlier,* largely based on the "energy security" argument. Today, the environmental community is skeptical of proposing 20 percent for 2020, despite colossal advances in wind energy and photovoltaics, and the national-security–crazed Congress will not call for a mere 10 percent contribution by 2020. But presidential candidate John Kerry said in a speech in January 2004, "I support a national goal of producing 20 percent of our electricity from renewable sources by 2020." *Plus ça change ....*

Not to end this chapter about US energy policy on a negative note, I must make reference to the amazing Solar Decathlon held on the Washington Mall, right in front of the Capitol, in 2002. It was sponsored by the DOE, BP Solar, and Astropower, the last wholly US-owned PV manufacturer (now GE Solar). Energy Secretary Spencer Abraham inaugurated the two-week event, which showcased 16 all-solarelectric full-size houses built by engineering

Neville Williams

departments at 16 US universities, from Hawaii to Puerto Rico. The universities competed to see which house was the most efficient and could provide the most power for everything from dishwashers and air conditioners to a small electric car that was being recharged outside

The 2002 Solar Decathlon on the Mall in Washington, DC.

each house. My alma mater, University of Colorado, won, and I was so proud of the enthusiastic young men and women who had devoted so much time and effort to designing, building, crating, shipping, and reassembling these high-tech houses on the mall. Seeing a whole solar community with thousands of square feet of glittering, blue, solar modules shining in the summer sunlight right in front of the US Capitol was an inspiring sight. This was "hope" on steroids.

At the Solar Decathlon's temporary "village," I spoke with David Garman, the Bush administration's assistant secretary for the solar office at DOE, now renamed the Office of Energy Efficiency and Renewable Energy. I reminisced

Neville Williams

University of Colorado students built this all-solarelectric house: first prize winner at the Solar Decathlon (2002).

about the early days at DOE, when Tom and Omi and Paul and the rest of us pressed the first solar campaign, and said we never would have dreamed an event like this could be held in Washington or sponsored by DOE. He smiled and said, "See what you started!"

# CHAPTER II

# A World of Darkness

The huge hippo herd was cavorting along the shores of the great Zambezi River in the late afternoon sunlight. Robin and I raised our sundowners to the glory of Africa, especially the Africa we'd once known, which was fast disappearing.

Robin Moser, a third-generation South African of Swiss-Irish descent, said, "We used to drive here from Jo'burg, to Vic Falls, to take the BOAC flying boat to London. There was no other air service, and this was the only river big enough for the four-engine amphibious plane to land. From here it would fly to Lake Tanganyika, then to Egypt and the Nile, then on to Europe."

I had my own memories of the old Africa. I had motorcycled from Cairo to Cape Town in 1962–63 and had fallen in love with the continent before most of it fell to war, famine, and corruption. I had met Robin quite by chance as we stood on the high, steel arch bridge, planned by Cecil John Rhodes and inaugurated by the son of Charles Darwin in 1905, which stretched across the Zambezi below Victoria Falls.

"You've come on that all the way from Cairo?" the young man had inquired, looking at my 250 cc, two-stroke "scrambler" motorcycle. Robin was visiting the falls with his family. We became friends, and he invited me to stay at his home when I later got to Johannesburg. We remained friends over the years and now, nearly 30 years later, I had come back to Africa for the first time and had looked him up. He immediately suggested we revisit Vic Falls together for a couple of days. We booked an Air Zimbabwe flight to Bulawayo, rented a car, and took rooms at that grand old Victorian pile, the Victoria Falls Hotel. Robin made sure I brought a jacket and tie for the dining room. "They keep the old traditions there," he said.

But this was not just a "looking back." I was looking to the future and what it could bring.

I believed South Africa had a vibrant future. I always knew it would settle the apartheid problem, and I had the feeling that 1990 would be *the* watershed year for profound changes. I wanted to be there when history was made.

Meanwhile, Zimbabwe, formerly Southern Rhodesia had emerged as the "hope of southern Africa." Robin wasn't convinced about this new emerging Africa run by the black majority, and this was clear when he saw the two African pilots enter the 737 cockpit at Jan Smuts Airport. He almost got off. On this particular flight the Boeing happened to make the hardest landing I had ever experienced, and I thought the wings would come off. "See," said Robin, "these people aren't ready."

I disagreed. "Ready or not, it's their country, and pretty soon they're going to be running *your* country, too."

I was looking at Africa now through very different eyes, at a very different time. The "winds of change," to borrow a phrase from Harold Macmillan's controversial 1960 speech before the South African parliament, had at long last blown new hope across the whole of southern Africa.

I wanted somehow to make a small contribution to this new "developing" third world, or what Paul Hawken, the author of *The Ecology of Commerce*, calls the "Two Thirds World." In the poorer countries, where the majority of the world's people live their short, happy, but often desperate lives, lay the great challenges for the planet. How would they develop? How would they acquire even a small measure of the lifestyles we took for granted? How could they have even such a basic thing as electricity without destroying the global environment?

Driving through the bleak Hwange coal country, Robin told me Zimbabwe and South Africa had more coal deposits than the entire world could ever burn. "Great," I thought, watching the huge Hwange power plants spewing clouds of gray smoke into the African skies, trying to envision what the air would be like when black people, as well as white people, were able to get electricity. I had read that only 2 percent of rural Zimbabwe, where 90 percent of the people lived, had electricity. I also learned that 98 percent of Zimbabwe's electricity was used for mining and industry, not for lighting homes, schools, or businesses. Kerosene and candles lit most of the homes in Africa — and most of the homes in the world. According to United Nations statistics at the time, two billion of the five billion people on the planet had no electricity. Now there are six billion people, and even more of them are without access to electricity.

All this coal! All these people without power! Everyone looking to a brighter future, figuratively and literally. And hydropower generated from the mighty Zambezi at Kariba, the enormous high dam I'd watched Italian workers constructing 30 years earlier, was not the answer because the electricity had to be distributed, which cost more than generating it. Besides, the massive hydroelectric projects of the 1960s had become their own environmental nightmares, displacing people, destroying ecosystems, and even altering microclimates. Dams and coal were definitely *not* the future.

At the Victoria Falls Hotel that night, I couldn't sleep. It wasn't the screech of the night insects that I could hear through the window — I had requested screens be put on so I could open the window to the cool night air. Nor was it the ceiling fan's breeze on my face. I was having a Big Thought.

"Why not use photovoltaics to bring basic electricity to these people?" I wondered, lying there in the Mzilikazi Suite, listening to the distant roar of the "Smoke That Thunders" outside my window. It was as if a light had gone on in that room, although it was as dark as only the African night can be.

Solar energy, which seemed to have little chance in America, as we'd learned the hard way at the Department of Energy, could be the best new energy source for the widely dispersed population in the developing world. And the environmental threat posed by the increased greenhouse gas emissions that would be released as the developing world developed could be partially avoided.

As a consultant to Solarex, then America's largest solar PV manufacturer, I'd learned that their biggest market for photovoltaic modules was South Africa, a fact they didn't advertise during the period of the anti-apartheid boycotts. PV was widely used for telecommunications, wireless telephony, radio and television broadcast translators, railroad signaling, lighting for remote post offices, and power for police stations. South Africa was the perfect crucible for this new technology because it was, as the brochures said, "a world in one country." It was the "first world" and the "third world" in one place, and a world of vast distances where whites all had electrical power, most blacks in rural areas didn't (blacks in the townships did), and white entrepreneurs and businesspeople saw a cost-effective "application" for solar electricity.*

The government was already putting solar energy to use in South Africa in a big way, and a few small Afrikaner-owned businesses were actually selling solar home-lighting systems to black farmers, who could afford them on "lay-by" plans if they didn't have the upfront cash.

Before returning to the United States, I flew to Cape Town. I had a hunch that something momentous was about to happen.

---

* I'm trying hard to avoid bureaucratic jargon like "applications" and "implementation" and "strategize" and all the strange words "institutional" policy writers lard their reports with — a language I had to learn in order to raise money in the nonprofit and "multilateral development agency" worlds.

Prime Minister F.W. de Klerk had hinted that some time that year Nelson Mandela would likely be released from prison, where he had proudly kept the dream of the African National Congress (ANC) alive since 1964. My white South African friends could not and would not believe it. Robin had driven me in his Rolls-Royce out to his weekend farm in the Magaliesburg. The transmission line to his country house bypassed thousands of black homesteads along the way. I tried to talk to him about Mandela, but he wouldn't have it. "No way they're going to be allowed to run this country," he said, echoing British prime minister Margaret Thatcher, who had said in 1987, "Anyone who thinks that the ANC is going to run the government of South Africa is living in Cloud Cuckooland."

Thieves kept breaking into Robin's weekend retreat to steal his stereo. "I just buy another one," he said, nonchalantly.

"How do they power them in their huts without electricity?" I asked.

"They hook them up to batteries. They steal those too."

"Why don't you just buy your African neighbors stereos and arrange for electric lines to their houses," I inquired. "You could afford it." He gave me a dour look. I suggested that solar photovoltaics might help.

Now, in Cape Town, I was watching a country in transition. I climbed atop the city market building at City Hall Square, which lay before the grand edifice designed by the prolific colonial architect Herbert Baker. Violence broke out. Sixty-nine people were shot, four killed, on that long hot February day under the searing Cape sun. I shared my binoculars with polite young black South Africans, who were having as hard a time as my friend Robin believing that what was about to happen would actually happen. But it *was* happening, right there before our eyes.

I could see them plainly through my binoculars, and suddenly everyone around me needed to have a look. Nelson Mandela, with his wife Winnie, appeared on the steps of City Hall before the mixed-race crowd of 80,000 expectant supporters who'd waited, as I had, almost six hours in the shadeless plaza. Mandela was free, and so was South Africa, and so was the human spirit that believed a better world could be made here and elsewhere! Maybe it wouldn't be, maybe it can't be, but it never hurts to believe, and symbols of that belief like Nelson Mandela are God's gift to mankind. The "winds of change" had finally blown clean through this beloved country. With the highest solar radiation in the world after Saudi Arabia and Australia, South Africa was now a country where I could legitimately pursue my own dream of helping to bring solar power and light to rural people still living in the dark.

Back in Chevy Chase, Maryland, I took over our dining room table to form an organization, or a company, that could promote solar power around the world and help bring it to those who needed and could use it. No solar company had yet made a profit, so that approach didn't bode well. I thought about setting up a "nonprofit," and that got the following response from my friends at Solarex: "Sure! We're a nonprofit, too. Your not-for-profit organization and our nonprofit company can work together to sell solar to all those folks in these countries who don't have any money. Great idea!"

Because I didn't know what I was getting into, because no one else was doing this (besides one Richard Hansen in the Dominican Republic, whom I'll get to shortly), because I simply had nothing more important to do, and because friends thought it was a great idea, I decided to set up a nonprofit organization to promote and finance solar rural electrification in the developing world. My old friend Oliver Davidson at the State Department's Office of Foreign Disaster Assistance thought it was "a good idea," and he had the international development credentials to back up his opinion. I had also consulted Dr. Peter G. Bourne, former special advisor to President Carter and assistant UN secretary-general, who strongly supported the idea of an "NGO" (nongovernmental organization) devoted exclusively to bringing solar power and light to the developing world.

I thought long and hard about a name, and in the early summer of 1990 I came up with Solar Electric Light Fund. I borrowed the name from

Thomas Edison's "Edison Electric Light Company," which pointedly revealed that the original electric power companies were set up to deliver one thing only: electric light for houses, business, and industry. They were all "light" or "electric lighting" or "illuminating" companies. I would just add the word "solar." The acronym was SELF, for energy SELF-sufficiency and SELF-reliance.

The Solar Electric Light Fund was officially launched on June 23, 1990. Its mission was to bring solarelectric lighting to rural people in the developing world who had no access to electricity. We called it an "electrifying idea."

Patti Forkan, my wife and executive vice president of the Humane Society of the United States, along with Ollie Davidson and I, formed the founding board of directors. I'd known Ollie since Ohio, and we had been in Vietnam together — I as a freelance correspondent, and Ollie as a civilian "district advisor" for the US Agency for International Development (USAID) in Hau Nghia Province near the Cambodian border. Ollie had first introduced me to exotic lands through his talks and slideshows in Ohio, where he presented the backpack travel adventures of his college years. Long before the world was overrun with "backpackers," there was Ollie, journeying through the Amazon or up the Nile with his army surplus rucksack. Before USAID hired him to join our misadventure in Vietnam, he would return to remote villages in West Africa and along the Amazon, where he'd earlier made friends during his backpacking days, to bring them medical supplies. No one I had ever met cared more for the common people in the poorer countries than Ollie, and I was honored that he not only thought the Solar Electric Light Fund was "a good idea," but that he agreed to be a director. Peter Bourne joined the advisory board and opened doors for us at the UN and United Nations Development Program (UNDP).

Paul Maycock, whom I'd worked with back at the DOE, also agreed to join the board. Having run the DOE's PV program for Carter, he now was regarded as the most knowledgeable expert on the worldwide photovoltaic industry. Paul told me, "As I get older, I find myself more motivated by the role photovoltaics can play in providing subsistence levels of electricity for the two billion people of the world who have no electricity. These people have little chance of ever seeing electricity in their homes, clinics, or schools." He understood the problem and the solution.

Now I had to raise money. Solarex, thanks to its CEO John Corsi, made the first grant of $5,000, even though John skeptically wondered how the hell I would ever figure out how to sell solar electricity to poor people in the developing world. "It's a huge market, but these people don't have any money," he reminded me one more time.

I didn't have any money either, not even a salary or an office. With the Solarex grant I bought my first computer, a Zenith laptop with an 800 kB memory, and began writing grant proposals. To my great surprise, a couple of small foundations specializing in "environmental issues" came through with support, and SELF was off and running. I opened a one-room office in a third-floor walkup on Connecticut Avenue, near Dupont Circle, and installed a cheap air conditioner. Tom Tatum from DOE days hosted a small fundraiser in the Hollywood Hills, film producers and B-list actors dropped by, and we raised $1,000. SELF maintained this link for the next six years, and some of Hollywood's most famous and committed environmentalists became regular supporters. Activist actor Ed Begely Jr. joined SELF's advisory board and hosted a fundraiser at his house, where I met the original "Tonight Show" host, Steve Allen, whom I'd grown up watching in the '50s when he was America's biggest TV celebrity. He was keenly interested in our plans to bring electricity to Chinese peasants, as he had traveled a great deal in China and had written a book on the country with his wife, actress Jayne Meadows.

Speaking of Hollywood, we later tried to reach actor Jack Nicholson after he did an interview for an environmental magazine in which he said the only hope for mankind was solar electricity. "Solar electricity, solar electricity, solar electricity," he was quoted as saying. "There, I've said it again." We could never get to him. Many years later, however, Larry Hagman, J.R. Ewing in the TV series "Dallas," joined SELF's board of directors.

Back in DC I bought a fax machine, which was a relatively new gadget in 1990. Before e-mail, fax machines shrank the world. This was the miracle device that allowed inexpensive communication beyond the normal borders of international telecommunication, and when messages began to arrive from Baluchistan (yes, Baluchistan — a province in southern Pakistan) and other exotic places, I felt like I'd been admitted to Marshall McLuhan's "global village."

McLuhan, the visionary Toronto professor who prepared us for a world made smaller by electronic communications, had inspired me more than I realized. I found my original paperback of his book *Understanding Media: The Extensions of Man*, published in 1964, in which I'd underlined that same year — the year I met him at the University of Colorado after a speech attended by exactly ten people — "Light is a non-specialist kind of energy or power that is identical with information and knowledge .... Grasp of this fact is indispensable to the understanding of the electronic age. *The electronic age is literally one of illumination*" (my italics). McLuhan had always been my prophet, but I didn't really know why until I caught up with him 30 years later.

More grants came in, and I began looking for countries where I could apply my concept of small-is-beautiful solar power for householders. Serendipitously, the countries found me. They usually reached me by fax, or responded by fax if I contacted them first by mail. The first was, in fact, the Island of Serendib, once known as Ceylon, which changed its name to Sri Lanka in the 1970s. The word "serendipity" was coined when an 18th-century English writer found himself in this island paradise below the equator. It means, according to modern dictionaries, "the faculty of making happy and unexpected discoveries by accident."

This was to be the story of SELF over the next seven years. Our quest to bring solar power and light to remote villages in 11 countries was serendipity squared. But before heading off to Sri Lanka, I needed to do a little more research.

While investigating the possibilities for this unique venture, I had arranged to meet with a young World Bank researcher, Michael Crossetti, at the Energy Sector Management Assistance Program (ESMAP). Not only was this my first encounter with the World Bank, about which you will hear more later, but it was also my initiation into the world of acronyms, around which the wheels of global development seem to turn. In Washington, people speak in acronyms, kind of like speaking in tongues.

Michael knew something about photovoltaics. He had published an ESMAP study on large-scale and mini-grid solar power installations in Pakistan. These were all multimillion-dollar projects featuring "central power arrays" of PV modules, which were connected to individual homes, or which fed power into the local electric grid.

The wholesale failure of these projects, described in the report, had given PV a bad name as a development tool.

More than anything, reading this study convinced me SELF was on the right track. Engineers had simply not been able to conceive of using PV modules in a small and decentralized way, house by house. Thus, the companies like Solarex, ARCO Solar in California, and Total Energie in France, a division of the oil company now known as Total-Fina-Elf, had foisted their highly engineered megaprojects using solar photovoltaics on the various development agencies, including the UNDP, USAID, and individual country donors like SIDA, DANIDA, NORAID, NOVIB, CIDA, and GTZ. I learned which acronym sponsored which project. All were disasters.

They failed, according to the ESMAP report, because people stole the copper wires that connected the houses to the PV arrays. And then people stole the PV modules because they wanted to put them on their own houses to charge batteries. They failed because the power in the batteries could not be "managed" and delivered equally to all the households, despite sophisticated electronics, which tended to break down in any case. Before any other power solutions were offered them, the local people were already using car batteries, charged once a week in town, to run their TVs. They also knew that *one* PV module could supply their house. Thus, a 200-module array, engineered to supply Western levels of electric service to 20 houses and to satisfy the design parameters of Western engineers, could in fact serve 200 houses, one by one, with the subsistence level of power the houses actually needed.

Engineers could not grasp this, and the procurement managers and project bidders were not interested in anything but large-scale, Western-style projects. In addition, none of the multinational suppliers of solar hardware for these projects ever bothered to train locals to maintain them. In most cases, once the construction engineers left, the solar power project would be a ruin within three to six months. And the only people to benefit were local householders, who acquired the modules for their own houses; thieves; and the solar companies, which had already cashed their checks and gone on to propose more such projects.

There were also two technical considerations that had to be addressed. First, solar PV produces only direct current (DC), and DC power, as Thomas Edison discovered, doesn't travel very far. Thus centralized PV systems are inefficient because they must send their power hundreds and even thousands of feet over very thick wires, resulting in large "line losses." Second, solar power can only work in the daytime, so power must be stored in batteries at night, and how do you control how much each home uses each night? Engineers have tried to solve this problem, but have always failed, for reasons more clearly laid out in Chapter 4.

Michael did not dismiss PV as a development tool, however, but strongly advocated stand-alone solar systems. "It is shown," he wrote, "that a decentralized approach for household PV systems in which individual households or buildings are powered by individual PV systems is less costly than a centralized approach in which a village is serviced by a single PV array and mini-distribution system." He based his conclusions on studies of demonstration PV projects in other countries, organized by the German and French governments.

Michael's report concluded: "PV is clearly the least-cost technology for off-grid village/household power supply." That's what I had thought, based on my own initial research, and that was exactly what we had set up SELF to do. I was delighted to see the experts concur, based on expensive, detailed analysis and evaluation by the World Bank.

This was not of interest to the engineers who dominated the solarelectric industry, however, since they had no way of selling their product to individual households; they wanted megaprojects, paid for up front by the United Nations and by various developing world governments. Remember, "These people don't have any money" had been their mantra. Environmentalist Barry Commoner had proposed in an article that if we distributed shiploads of PV cells, instead of barrels of aid money, to governments of poor countries, they could assemble the cells themselves into PV modules (adding local value) and distribute them on a heavily subsidized basis to power their whole rural economies. At the same time, this would keep the aid money at home, purchasing state-of-the-art high-tech solar cells from US manufacturers. The idea was too wonderfully pure and simple to ever become reality. This technology attracted its share of dreams and dreamers, that was for sure. (A point of clarification here: A solar "cell" is 3 to 5 inches wide. Placed together, 6 to 36 solar cells become a "module." Two or three modules form a "solar panel," and three or four modules become a "solar array." See Chapter 4.)

In India, the government decided not to wait for Barry Commoner's shiploads of cells and set up solar-cell manufacturing plants itself in the late 1980s, but the engineers and politicos in charge were not disposed to distribute their product to individual households, preferring the same megaprojects and centralized solutions that Michael's report had so clearly criticized. (I met some of these engineers and politicians later — see Chapter 6.)

Michael told me that he knew of one engineer who understood the basic simplicity of stand-alone DC power produced by individual solar PV modules, installed house by house, each with its own battery, avoiding transmission costs and power losses. I still needed "proof of concept," as they say, as I had not yet seen an actual house powered by its own small solar module. I had only envisioned it, like the dreamer I was.

The engineer ESMAP referred me to was Richard Hansen, who had been working in the Dominican Republic for the past four years, installing solar home systems (SHSs) through his Massachusetts-based nonprofit group, Enersol. A former Westinghouse engineer, Richard had fallen in love with the "DR" during a vacation there and returned one day to marry a Dominican. As an engineer who had studied renewable energy systems, he saw the potential for solar PV as a useful technology to make people's lives better. Since the DR had a pitiful electric utility that could not deliver power beyond the main towns, and that suffered blackouts much of the time when the government could not afford to buy the imported oil needed to run the generators, Richard decided to do something about it.

He brought in several solar modules, hooked them up to batteries in a house he rented, and invited the locals to see it. The ability to have bright electric lights at night, watch a small TV, and listen to Dominican salsa on their radios by day created an instant demand for solar systems. Some families spent $10 a month on radio batteries in order to listen to music, which to Dominicans is almost as important as food. They spent even more money on car batteries, which they hooked up to small black-and white TV sets, and an additional sum on kerosene for lighting. For the same money they could buy an SHS, or they could buy it on monthly installment payments if credit were available.

Richard formed a US nonprofit, found a few private donors, got a $2,000 grant from USAID, and set up an office outside Puerto Plata on the north side of the island. The Martinez family bought the first system on sight with a down payment and a three-year loan from the revolving fund Richard set up. Solar finance was born. By 1990, Enersol's community solar program had financed over 1,000 customers. Over the years, Richard hired half a dozen ex-Peace Corps volunteers to help in his quest.

I contacted Richard and his assistant, Phil Covell, and was cautiously offered a chance to visit Puerto Plata provided I took care not to "disrupt the program," which foreign visitors were often accused of doing. American Airlines has flights directly into Puerto Plata from Miami, and I arrived at the two-room airport in October 1990. Richard collected me in his jeep and we immediately went searching for gasoline. A lack of foreign currency prevented the government from paying tanker captains for their oil shipments, so the oil ships sat offshore, waiting for dollars. Naturally, there was no electricity either, since the power plants ran on diesel fuel.

The DR, with seven million people, is a microcosm of the problems of the Two Thirds World: too many people, no resources, terrible corruption (in 1990, President Joaquín Balaguer was nearly 90, blind, and useless), worthless currency, horrible climatic conditions — there are usually hurricanes every few years that set the island back decades. I'd traveled the world and been all through Mexico and Central America, but I had never seen poverty at the level it exists in the DR, a country one hour's flying time from Miami, with more people than Switzerland or Greece (not counting the million Dominicans who — no wonder — live in New York City).

Nonetheless, the rural areas were gorgeous and unspoiled, and they offered promise to farmers who wanted to work. But there was no electricity. Many farmers earned a decent income and built decent houses, which they painted in bright colors. These were Enersol's customers.

"We let the sun distribute the power, not copper wires," Richard pointed out. He took me to visit two dozen houses up a remote valley reaching into the tropical hills. Every one had music playing, many had bought televisions, and everyone was happy to have electric lights at night, including the owners of the small shops along the way. Nothing is darker than a rural village without electricity, and one 12-watt DC fluorescent bulb (equal to a 60-watt AC bulb) looks like a prison-yard spotlight in that penetrating darkness.

Some of the wealthier families had bought DC juice blenders and even large stereo sets. Everywhere children were smiling, and the families invited us in for fruit or a soft drink. At the end of the day, Richard took me back to his own house, which he'd built with his own hands, a two-story concrete and cement block affair with wraparound verandah and solar-powered everything: hot water, computers, TV, radios, stereo, water pumped from his own well, and lights everywhere. He had a kerosene refrigerator because DC fridges were still expensive.

We sat in his office and studied revolving loan funds. This was the key. Without credit of some sort, which we in America took for granted, people could not afford these solarelectric systems. Richard had pioneered the idea of revolving credit funds, but he had not had much success finding the funds to revolve. He wasn't a bank. When he realized that a $100,000 fund loaned out at 10 percent on three-year terms would only finance 800 homes over five years, he decided to try a different approach — selling the service, like a utility, instead of the actual hardware. But I'm getting ahead of the story.

Meeting Richard further convinced me that SELF was on the right track. Back in Washington, as a way to say thanks for the inspiration he provided me, I found some financial support for him from USAID and Winrock International, the global agricultural NGO founded by Winthrop Rockefeller, former governor of Arkansas (and one of the Rockefeller Brothers, the five sons of John D. Rockefeller Jr.). Then I set about raising more money for SELF so we could embark on our first project that, serendipitously, had emerged from Sri Lanka.

I had been to Sri Lanka in 1972, when it was still called Ceylon and when it was still a true paradise, before the ethnic and political civil wars started, before the devastating deforestation that altered the monsoon patterns and caused droughts and killer floods, and well before the 2004 earthquake-generated tsunami that killed 35,000 people. Sri Lanka, 270 miles long and 140 miles wide, is one of the most beautiful places on earth. It has thick jungles and open plains, meandering rivers and high mountains, ancient cities and spectacular ruins, wild elephants and tame ones, and lots of wildlife. Its 18 million people are the best educated in the developing world. The great majority are Singhalese, mostly Theravada Buddhist, while 15 percent are Hindu Tamils, and there are small populations of Muslims and Christians. Ten percent of the population speaks English.

Sri Lanka was among the first of the British colonies to achieve independence, in 1948, and it was the first former colony to give women the vote or to have a woman as prime minister. It can also boast of producing the first female suicide bomber, who traveled to India and blew up Prime Minister Rajiv Gandhi in 1991. This happened on my return trip to Sri Lanka.

In Moratuwa, 30 kilometers south of Colombo, I met our project partner-to-be, Dr. A.T. Ariyaratne, president of the Sarvodaya Shramadana Movement, to whom I had been introduced by the W. Alton Jones Foundation in Charlottesville, Virginia, which offered funds to SELF to set up a pilot solar program. (W. Alton Jones was an oilman whose heirs were trying to do good things with the family money by supporting unique environmental initiatives, including our plan to bring electricity to poor people without producing more greenhouse gases.) Despite my earlier travels in Sri Lanka, I had not heard of Sarvodaya, one of the largest NGOs in South Asia. Founded in the 1950s by Dr. Ariyaratne, or "Ari" as everyone called him, Sarvodaya — which means "universal awakening" — was built from the grass roots up, and it focused from the beginning on what Ari called "people-centered development." From a village-based "shramadana" movement (meaning "sharing labor and energy"), Sarvodaya had grown to a membership of over two million people, with development activities in 8,000 villages. To many people, Dr. Ari was the contemporary Gandhi of South Asia, equally admired in India and other countries in the region.

Dr. Ari welcomed me with a great hug, despite his diminutive stature. Then he lit up the first of many cigarettes. We entered his inner chambers at Sarvodaya headquarters, a sprawling compound at Morotuwa. On the wall were photos of Dr. Ari with numerous heads of state and at global religious and development conferences. I was immediately struck, not by how famous he was, but how humble: a more down-to-earth, easygoing person I couldn't have imagined. But his energy was palpable, as was his charisma.

"So we're going to do a solar project together!" he said. "I've been a long believer in solar energy and new technologies." He presented me with copies of a half dozen published books about rural development in which he offered his practical views on how to lift people from poverty and illiteracy. He believed in practice, not theory, unlike the economists who staffed most of the development agencies. Dr. Ari confounded his "bilateral" (single country) donors by having as his only "theory" the belief that poor villagers, if given the chance and the tools, could lift themselves out of poverty.

One of his tools was self-financing, and he had established Sarvodaya Economic Enterprises Development Services (SEEDS), an island-wide credit union that made small loans to hundreds of thousands of families who had never been inside a bank. This was long before "microcredit" became fashionable. We were going to set up a "solar fund" within SEEDS that could self-finance the purchase of SHSs.

First, Dr. Ari drove me, in his Mitsubishi Pajero, on a dark journey through the jungles to visit remote communities eking out a subsistence living growing coconuts, cashews, and rice. It was nighttime, and there were no lights anywhere. Small bottle lamps illuminated the rough wattle-and-daub houses with their flammable thatch roofs. People were thrilled to have a foreign visitor and proud to show me their simple lamps and lanterns.

Sarvodaya's Dr. A.T. Ariyaratne ceremoniously turns on an electric light switch at SELF's first solar-powered schoolhouse in Sri Lanka.

Dr. Arthur C. Clarke, author of *2001: A Space Odyssey*, who had lived in Colombo for 30 years, later told me that 10,000 people a year were maimed and disfigured by kerosene lamp fires in Sri Lanka. This was why he was such a big supporter of solar electricity. He had written, "The village in the jungle can leap-frog into the modern age, thanks to solar electric conversion devices."

I was confronting directly the poverty of the Two Thirds World. "Can we help these people with solar power?" Dr. Ari asked me. I tried to figure out how much money these simple villagers had, or earned, but that was not an easy question to ask and, if asked, was not likely to get an honest answer. However, it was clear that without some kind of financing assistance, and without low-cost but reliable SHSs, these people had no hope of moving up to electric lights. And it was certain the government was not going to bring electricity to them. Without solar lighting, they would remain in the dark all of their lives.

When I visited the offices of the Ceylon Electricity Board, I learned that 70 percent of Sri Lankans remained without power, and the barriers to their getting a grid hookup were money and generating capacity, which was limited by how much rainfall filled up the hydro dams. The electricity board officials told me a coal-fired generating plant was planned, the island's first, which would have to *import* coal from Indonesia. No one imports coal. But Sri Lanka had only hydroelectric plants that used a so-called renewable resource that was not always sure to be renewed.

There, at the front lines of rural development, stood Dr. Ari, seeking solutions and looking at solar photovoltaics for the first time. Over the next five years we did build a successful solar program in Sri Lanka, which became the base for the largest solar project financed by the World Bank. What I learned in Sri Lanka, after endless meetings with World Bank economists, power company officials and planners, village-based Buddhist leaders, and local environmentalists, was that we could deliver an SHS for less than it cost to run an electric line 100 meters from the nearest power main. And this didn't include the cost of transmission or generating the power, just the hookup and house wiring. The sun is, indeed, a better distributor of power than copper wires, especially when houses are widely dispersed, as they are in rural Sri Lanka and in most developing countries. Sri Lanka became our solar learning lab, and today it leads the world in solar installations. One day it may become the world's first solar-powered island.

The Solar Electric Light Fund was now off and running. I still had no staff, but grant funding was coming in for projects in China, South Africa, Nepal, Vietnam, and an additional project in Sri Lanka. For between $30,000 and $50,000 in project grants, it was possible to launch a pilot project of a few hundred households in these countries. "Core support" grants covered travel expenses, office rent, and my modest salary. This allowed SELF to use project grant money in its entirety for each project, without deducting administrative costs. I intended to give the environmental grant makers who had taken a chance on SELF the biggest bang for their buck. They were used to seeing enormous sums absorbed by large "program" and administration

staffs at many of the nonprofits based in the nation's capital, with little left over for actual "program activities" or real projects.

SELF was like a guerilla nonprofit — fast on its feet, hard to see (since all the projects were at least 10,000 miles away),

Author assists a Sri Lankan boy as he switches on an electric light for the first time.

cost-effective, and decisive. We got results because of all the exceptionally honest, hardworking, and dedicated people I was able to find in my travels to a dozen countries.

I learned that many nonprofit organizations in Washington spent their grant money on policy studies, research, travel, high salaries, glossy reports and brochures, and chrome-plated overhead (just visit some of the nonprofit headquarters in DC and you'll see what I mean). Many were staffed by over-paid academics and bright, young, underpaid youngsters. Home to dedicated policy wonks, many nonprofits did excellent policy work, influenced law-makers, and educated the public on myriad issues, but few could actually point to a *visible* project or *concrete* results. Maybe they *were* making a differ-ence — I could never tell. In any case, these organizations all took themselves very seriously, and there seemed to be no shortage of philanthropic funding for them.

Maybe it was because I was older or just a simpler person, who had not grown up in Washington's heady, self-important, deadly serious environ-ment, but I had little patience with the way some things were done in the capital's nonprofit community — especially with regard to international "programs." I wanted simple, practical, working *solutions*. I also wanted to see hands-on, tangible results; something that directly affected people; some-thing I could photograph! It didn't seem we were really making a difference unless human beings benefited *directly*. SELF soon acquired a reputation for taking action and getting results with very little money. I learned that I could establish, through actual example, a national solar-electrification program in a country with less than $50,000. That amount would barely pay for one World Bank consulting mission or the annual in-country office rent of a bilateral donor. It wouldn't pay a quarter of the salary and expenses of one USAID officer.

I was soon having a wonderful time, meeting so many terrifically inter-esting and entertaining people around the world as I traveled to their countries. I was invited by the UN Development Program to head a UNDP-World Bank mission to Zimbabwe to design a national solar program. I was invited by Will Cawood in South Africa to come and electrify a village in the Valley of a Thousand Hills. I was invited to China by Mao Yinqui, whom I met at a solar energy conference in Denver in 1992. A group of Nepalese professionals invited me to Kathmandu to help them launch a rural solar program. And the Vietnam Women's Union, an 11-million–member organization, invited me to help them bring solar power and light to their poorest and remotest constituents.

These projects — covered in the following chapters — all meant years of hard work, but they all got "implemented" successfully, to use a word popular among development types. Of the 12 countries we worked in, it was only in Zimbabwe that SELF was not able to actually implement the project because we had designed it. (Under UN rules, the designer cannot implement or manage a project.) I learned a great deal while designing this project, and I was pleased when it was passed and budgeted for $10 million at a meeting at UNDP headquarters on 42nd Street in New York, where I had to defend my "planning document" before a dozen UN and UNDP officials. Under the direction of Gibson Mandishona, whom I hired over tea at the Monomotapa Hotel in Harare, the Zimbabwe project eventually spawned 40 small African-owned solar companies, which together managed to install over 11,000 solar home-lighting systems around the country, financed by local banks. At a solar conference at the Harare Sheraton in 1995, Robert Mugabe, Zimbabwe's only leader since independence, gave the most powerful speech I had ever heard in support of using solar power to electrify the country's rural population. How sad it is to watch Zimbabwe implode seven years later under Mugabe's subsequent monstrous and demented rule.

To the south, after speaking the same year at a solar energy conference in Swaziland, where women carried car batteries on their heads to market for recharging, I made plans with Will Cawood, a South African solar engineer, to electrify an off-grid community in KwaZulu Natal. Mapaphethe was a Zulu village of a thousand households not 20 minutes from the Durban-Pietermaritzburg freeway, but it was outside Durban Electricity's reach and not in the rural electrification plan of Eskom, South Africa's huge power utility.

Its young chief, who as a Zulu "Nkosi" held the power of life and death over his subjects, was distressed that he could not get his community electrified, even though high-tension power lines were visible in the distance. With the help of the KwaZulu Finance Corporation, which was willing to make bank loans to these rural people, most of whom had city jobs and a modest income, we were able to electrify 150 homes as an example of how people could have electricity *now* instead of waiting for the government-owned utility to bring it to them.

Mapaphethe became a national "exemplar" project and was visited many times by the energy ministers of South Africa and surrounding countries. The grant money was provided by the US Department of Energy. The entire project cost $45,000, which covered $15,000 in expenses, including salaries for Will and his technicians, as well as a local training course, plus $30,000 for a loan guarantee for the finance company. The funds remained on deposit for years, earning interest that helped Will cover maintenance expenses. People bought their own systems at cost, supplied by Siemens Solar South Africa and installed by village technicians. This was the perfect solution: use donor funds for bank guarantees so that local lenders could offer loans for a consumer product that they normally wouldn't lend money on.

I was happy to be back in South Africa. Since Nelson Mandela had been elected president it was a much happier place — people smiled at you as they did in the rest of Africa, but had seldom done in South Africa under apartheid.

Our solar customers smiled even more, and I never ceased being amazed when I was invited inside their round, earthen "rondavels" (huts) to find that nearly everyone had a stereo system with large amplifiers and huge speakers, all running on solar power. Was this poverty? They had electricity and enough income to meet their needs, for the most part, which they supplemented with lush vegetable gardens, and they enjoyed the most beautiful climate on earth, with views to die for. I'd seen worse.

China called. In 1993 I followed up Mrs. Mao's invitation to visit the Gansu Natural Energy Research Institute (GNERI), which was western China's leading institution fostering all kinds of solar energy. Under the direction of Professor Wang Anhua, a true solar pioneer, GNERI had developed parabolic solar cookers, solar water heaters, and large photovoltaic systems and had instituted passive solar design for government buildings. GNERI's works were scattered from Tibet to Sichuan to Chinese Mongolia, and several hundred thousand solar cookers dotted the villages of Gansu, China's second-poorest province, where they glinted in the sun like rhinestones against the brown landscape. They could boil a kettle of water in 60 seconds with the sun's focused light; fortunately the sun moved across the sky a degree or two before it could melt the kettle.

Will Catwood

There were solar water heaters on the roofs of nearly every house in thousands of villages in China's cold north. Remote forest ranger offices and road crew barracks now had electric lights and TV sets thanks to Professor Wang's solar designs and relentless promotion of GNERI's solar technologies. GNERI

Female Zulu technicians installing a solar module atop a house in KwaZulu-Natal, for SELF in South Africa.

sold most of its PV systems to government agencies for use in these various outposts, and Professor Wang installed them.

SELF hired the professor to install 100 experimental solar home-lighting systems manufactured by GNERI in MaGiaCha, a remote village in Tongwei County. I was the first Westerner the people there had ever seen, and I had to arrange special permits to travel along this section of the Silk Road, which may not have been visited by a white person since Marco Polo (I'm not exaggerating). Mao suits were still common attire in Tongwei County, except for one young peasant girl who dressed up in a bright red jacket and donned high heels for a photograph. My first trip to China was in 1979, and when I visited communes then, peasants were proud to show me their alarm clock and thermos bottle and *maybe* a bicycle, their only possessions. Now, even in this poorest of regions, where people had starved to death during the Great Leap Forward (a failed campaign to industrialize the country in the late 1950s), nearly every household managed to buy a television set once it had acquired a solarelectric system.

With funds from the Rockefeller Foundation, Professor Wang and SELF formed the Gansu PV Company to manufacture the small household "plug-and-play" solarelectric systems that mystified visiting electrical engineers because of their amazing efficiency and reliability. Professor Wang and his son Wang Yu designed all the electronics themselves. SELF was now a 49 percent owner of the first "Sino-American" joint-venture solar company in China. And MaGiaCha became famous as the first all-solar village in the country. It was featured on TV and in the papers, and all the Communist Party officials, from the provincial vice governor on down to those at the county and district level, threw great banquets in SELF's honor and made their many pilgrimages to the remote commune nestled in the spectacular Loess hills of Gansu.

Not everything turned out as we planned, but we were "learning by doing," which became SELF's mantra, along with Goethe's dictate that "action begets action." The philanthropic foundation community seemed to agree, since the grants kept coming, project by project, as well as funds for overhead in Washington, and I kept traveling and organizing more projects.

The Rockefeller Foundation, the Rockefeller Brothers Fund, and anonymous members of the Rockefeller family became big supporters of SELF's activities. I loved taking old oil money to develop carbon-free sources of power and light that didn't require reliance on fossil fuels.

The most important lesson during these years was our discovery that, in general, rural people are excellent credit risks and will do anything to make their monthly installment or bank loan payment so they don't lose the electricity they have come to rely on. They also hate debt. For exactly the same reasons, the American electric power cooperative movement was able to finance itself in the 1930s.

Soon I was able to hire an assistant, Ben Cook, who was right out of the University of Virginia, where he'd earned a combined degree in economics and physics, studying solar electricity specifically. Like so many well-educated, bright, and eager young people who grew up in Washington and took entry-level jobs in their home city, he had high expectations and was ready to take over SELF within his first year. I didn't let him, but he did end up running SELCO eight years later!

I also hired Robert Freling, an energetic global environmental gadabout. Bob was a Yale graduate in Russian language studies who spoke fluent Chinese and could converse in Spanish, French, German, Indonesian, and Portuguese, his favorite language. I thought this would be useful. Bob eventually became executive director of SELF and runs it to this day.

Now there were three of us at SELF, plus a Sri Lankan administrative assistant and a roster of consultants, so we could manage all the projects I'd started and launch even more. It was about this time that SELF outgrew its modest office on Connecticut Avenue and moved around the corner to a brownstone on 20ᵗʰ Street.

The next stop was Vietnam. It would be my first visit to that country since covering the war as a correspondent 24 years earlier. Would the Vietnamese welcome an American coming to "do good" after our war machine had killed nearly a million of their citizens — soldiers and civilians — defoliated and poisoned their forests, and dropped more bombs on their country, north and south, than were dropped on all fronts by all sides in all of World War II? I was about to find out.

# Chapter iii

# Illuminating Adventures

From the Air Vietnam airbus you could still see the bomb craters that dotted this land of green paddy fields like a case of acne. America actually dropped three times more bombs on *South* Vietnam, our ally, than on the north; even here in the green Red River delta near Hanoi, the scars were still visible. Farmers had tried to fill them in, used them as ready-made fish ponds, or just left them alone.

The war had ended two decades earlier, and by 1993 half the population of this lush, friendly, somewhat laidback country was under 25 years old. Young people had little knowledge of the war, which we were still fighting in America (note the POW/MIA flag that flies atop the US Capitol to this day). Because we couldn't get Vietnam out of our system, and because the Vietnamese, for some reason, *loved* Americans, there would always be a strange bond between our nations. Forgive and forget was their motto, while we were incapable of forgetting and continued to rub salt in the wounds of our first lost war.

I wasn't here to make amends for our war crimes against this harmless land, which never threatened the United States in any way — and never could have. I was here because a SELF consultant had arranged for me to meet with the president of the Vietnam Women's Union (VWU), who was also the second-highest ranking woman in the politburo in Hanoi, after Madame Binh, who had been a senior official of the Vietcong. I was here because 70 percent of Vietnam's rural people had no electricity — some six million households.

Madame My Hoa greeted me warily in her formal receiving room at the five-story mirrored glass headquarters of the VWU in downtown Hanoi. Short and plump, she exuded power and importance. We were contemporaries it turned out, the same age. She was from the south and had fought the Americans in the Mekong Delta. Now she headed up the largest semi-independent organization in the country, a huge, highly centralized, social service bureaucracy representing 11 million women.

In Vietnam, women hold a great deal of power within the family and handle the family's finances. As a result, since Ho Chi Minh granted them equal rights in 1945, they also hold a lot of power in local government and are strong players in business as Vietnam heads down the capitalist road. However, only two women, Mme My Hoa and Mme Binh, were serving in the highest echelons of the central government in the early 1990s.

I explained to Mme My Hoa how SELF worked: we provided "solar seed" funds to buy 100 or 200 solar home systems (SHSs), which we would install in households where someone — in this case it was the women — signed an agreement to pay 10 percent down and monthly installments over three to five years. We then advised our project partner how to set up and manage a revolving fund. The partner was expected to collect the money, which would be used for the purchase of additional systems (which SELF would procure for them at cost) and to pay for ongoing maintenance. SELF would not be paid back; this was a grant. SELF would also cover its own project overhead, and it was expected the VWU would do the same, which it managed to do successfully by attracting other donors, so as not to dip into the loan fund. If the pilot succeeded, SELF would seek to raise additional funds to purchase even more systems, with the hope that the project would eventually become SELF-sufficient.

Although SELF subsidized the solar systems to varying degrees by not tacking on a margin of profit, the basic *modus operandi* of our approach was to get people to pay for the systems. We didn't believe in giving them away. In a country where people expected to get electricity for free from the government, it would be an achievement to get them to pay anything at all for power. At the same time, if we did not collect the full cost of hardware plus an operating margin, or did not charge sufficient commercial interest rates on the solar loans, we would be sowing the seeds of the "solar seed" project's own destruction. Revolving funds will cease revolving if sufficient interest is not charged, and SELF would not always be there in future to supply low-cost systems with no markup. As I have said, we were learning by doing.

In 1994 I returned to Vietnam, the second of many trips to the country throughout the 1990s, to sign an agreement with the VWU, which turned out to be the best partner SELF ever had. The women took to the nonprofit solar business like old hands, flawlessly organizing our joint solar-electrification projects in remote Mekong Delta hamlets and elsewhere in central and north Vietnam. Officials in the ministry of industry, in charge of power and energy for Vietnam, began inquiring why SELF wasn't working with *them*. "Rural electrification isn't the job of the Women's Union, it's our job," they told both Mme My Hoa and me. We heard the same complaint from Electricity of Vietnam (see Chapter 8).

Once the VWU agreed to be our partner, I was able to raise funds from Robert Wallace, who ran a family foundation in Washington called the Wallace Global Fund. The family fortune came from a patent on hybrid corn, and Bob had long been interested in agricultural issues in the developing world. He was the son of Henry Wallace, President Franklin Roosevelt's vice president, former secretary of agriculture, and Progressive Party candidate for president in 1948, and as such he perpetuated a family tradition of progressive thinking and activism. Normally, foundations make their decisions in secret and rarely interview the applicants for grants, but Bob wanted to see me. I'd asked for $100,000 for the Vietnam project. "I don't want to just do projects," he told me. "I want to change policy at the national level." I said I thought our project could do that, as we were already affecting policy in a number of countries. I talked and talked and presented documents, photographs, and lots of boring facts and figures. My consultant, Mark, who had helped open doors in Vietnam, came with me to the meeting and told me afterward, "You talk too much and shuffle too many papers and documents during a meeting. Too much information."

"We'll see if I was effective if we get the grant," I told him, somewhat irritated at his criticism. We got the grant. And Bob came through with several more after I did a lot more talking and shuffling and explaining how his funding was being used effectively. His death in 2003 was a great loss to the environmental movement.

SELF soon acquired the support of the Rockefeller Brothers Fund (RBF — not be confused with the much older, much larger, and more famous Rockefeller Foundation) thanks to two young program officers, Michael Northrop and Peter Riggs. I never dreamed I would get a hearing at this elite philanthropic institution, presided over by David Rockefeller, former chief of Chase Manhattan Bank. Michael read the material I sent him and agreed to meet me in New York, and again I made my talkative pitch on behalf of two billion people living without electric light, showed photographs, and presented documents and reports. He and his colleague Peter, who took a keen interest in what we were doing in Vietnam, a country he visited from time to time, sent me a letter saying they had approved the first of several grants for Vietnam. They also funded SELF's activities in India (see chapter 6). Every $50,000 check that fell out of an envelope, and which I rushed down to the bank to deposit, meant another 125 families would be able to purchase solar-electric systems. And the money would replenish itself, revolve several times, and finance more systems.

Perhaps SELF's biggest attraction was that we were not giving away solar electricity, but were putting in place, at the local level, a SELF-reliant means for rural people to literally take power into their own hands. I believed, perhaps naively, that people in these countries should be given a chance to lift themselves up through their own efforts. Acquiring their own source of household power for lights and TV and radio, instead of passively waiting for the government to bring it to them, was a start. Solar technology, which had become affordable in the early 1990s, could do that. This message appealed to grant makers, international NGOs, local development organizations, and community groups. I even sold the idea to the World Bank, the ramifications of which you will come to understand in future chapters.

By the mid-1990s, in my working trips to nearly a dozen developing countries, I had to learn how to navigate the shifting and mysterious seas of "international development." Who did what? How did they do it? Who paid for it? I knew something of the UNDP from my experience in Zimbabwe, but I didn't understand the World Bank, so I set out to learn.

In response to environmental concerns expressed by Washington's legion of environmental activists, the Bank was taking a look at renewable energy, which people like Michael Crossetti at the Bank's Energy Sector Management Assistance Program (ESMAP) were recommending it do. The Bank knew the UN and various bilateral agencies had been funding solar water-pumping systems for years in the Sahel (North Africa), along with the misconceived centralized PV generating "plants" that I mentioned in Chapter 2. I had been hounding everyone I could meet at the Bank's various departments — environment, energy, industry, country desks — to consider the idea of "solar rural electrification." I explained we were already doing it, but only on a pilot basis. Unless governments got behind it, with favorable policies and money, it wasn't going to solve the rural electricity problem, even though it was the least costly solution and didn't produce greenhouse emissions. Most governments were in thrall to the World Bank on electric power issues because they needed to borrow the Bank's money for energy development. And the Bank officials I met made it clear they were not going to fund any more conventional rural electrification programs anywhere because it wasn't "economical." SELF was offering an economic alternative — solar electricity.

Let me say right here that, despite SELF's lobbying efforts at the World Bank and later at the International Finance Corporation (IFC — the private-sector lending arm of the World Bank, located in its very own grand building on Pennsylvania Avenue), SELF was not a policy advocacy group. Our strategy was to set up "replicable" projects in as many countries as possible, prove the concept, and attract more private philanthropy to underwrite our very focused activity. Successful pilot projects would speak for themselves far better than our lobbying and advocacy could do. We were not waiting around for the World Bank, or even governments, to fund our efforts. In any case, the World Bank only finances governments, not NGOs. I did not approach the World Bank with the idea of convincing it to stop funding big hydroelectric dams and fossil-fueled power plants in favor of projects that didn't damage the environment. I went downtown to see the Bank because *that's where the money was*, and because governments needed money to pursue any sort of energy development in their countries. After SELF and other solar activists succeeded in getting the Bank to open its energy lending to renewable energy, we became the intermediaries that linked the Bank's lending policies to requests for alternative energy programs by national governments.

I almost always got a polite hearing, especially from people like Chaz Feinstein, one of the new managers of the newly formed, $5-billion Global Environment Facility (GEF), whose office I invaded in 1990 while he was still unpacking his computer. Chaz had come from ESMAP as an energy analyst and economist, and he had been a big fan of solar PV. He opened numerous doors at the Bank, the GEF, and the IFC, and over the years money did begin to flow toward the solar solution. There was also Loretta Schaeffer, who came out of the policy side of the Bank and was appointed to head up a new department, the Asia Alternative Energy Unit, focusing on alternative energy. I barged in on her, as well, to announce what SELF was doing in several countries, including Sri Lanka. It wasn't long before I was heading a World Bank mission to that country jointly with Dr. Anil Cabraal, himself Sri Lankan-born and the World Bank's leading expert on photovoltaics, along with a couple of technical people (electricity and finance).

I was somewhat conflicted about the Sri Lanka mission in 1993 inasmuch as I was being paid by the World Bank to do an "assessment and evaluation" of SELF's solar rural electrification projects in Sri Lanka, which I had set up. But Loretta didn't seem to mind, and we needed the Bank, which had more money than all the foundations in America combined, and which allocated most of it to energy in the developing world. We wanted to divert a wee bit of it to clean, small-is-beautiful, energy solutions and to Dr. Ari's "people-centered development."

I took World Bank mission members around the country to meet Sarvodaya people, to see actual solar installations, and to meet happy householders who would invite us in for the customary round of tea and coconut juice. I arranged meetings with the key ministers, as well as with Ceylon Electricity and the heads of the National Development Bank and the Development Finance Corporation of Ceylon (DFCC). I finally convinced Loretta's energy group at the Bank — by then called The Solar Initiative — to finance the purchase of SHSs in Sri Lanka through these nationally chartered development banks and sundry microfinance institutions — i.e., Sarvodaya's SEEDS — *instead* of through the Ceylon Electricity Board.

This was a new twist for the Bank, as it was accustomed to financing large-scale generating plants directly, usually in partnership with the national electric utility. It had never financed banks to make solar loans. But after five years and hundreds of thousands of dollars worth of consulting "missions," we were successful. In 1998 the World Bank concluded a $32 million deal with the government of Sri Lanka and, ultimately, Savodaya, and the project began. At last there was enough money to make a difference, even if it was only an infinitesimal fraction of the Bank's annual energy loan portfolio.

By the time the first money flowed, Loretta had retired, Anil had moved on to other projects, and the numerous World Bank consultants assigned to the program had never heard of SELF or the story of how Sarvodaya and SoLanka, a solar NGO I co-founded, got into the solar business. The World Bank now took credit for everything.

Meanwhile, back in Washington, the World Bank was celebrating its 50th year, while environmentalists launched the "Fifty Years is Enough" campaign, suggesting that the best gift the Bank could give the world was to go out of business.

During the previous 50 years the gap between rich and poor had grown dramatically, and the Two Thirds World had gone deeply into debt on a borrowing binge that lined the pockets of corrupt ruling elites and kept the Bank's 7,000 highly paid economists gainfully employed and traveling happily on their "missions." Many of these economists worked on "structural adjustment programs," which involved convincing governments to cut social programs and development projects while opening their borders to foreign capital so that investors could exploit cheap local labor for export production and so that the subsequently "adjusted" country could pay back its suffocating debt. As Paolo Lugari, urban environmentalist and founder of Gaviotas, a sustainable tropical community in Colombia, once said, "Who creates poverty? Economists!"

A few years earlier, Bruce Rich had published *Mortgaging The Earth: The World Bank, Environmental Impoverishment, And The Crisis of Development*, which took the Bank to task in an excoriating, heavily researched diatribe that is still worth reading a decade later since, depressingly, so little has changed. Conservatives were agitating for change as well as progressives. Ian Vasquez, an economist at the Cato Institute and author of *Perpetuating Poverty*, said, "The Bank is not reformable. It should be abolished altogether." The *Washington Post* paraphrased bank critics who were "painting a grim picture of a multinational bureaucracy run amok, a secretive, bloated institution run by overpaid, jargon-babbling technocrats more interested in defending turf than tending to the poor."

Longtime activist Keven Danaher also wrote a book, *Ten Reasons to Abolish the IMF & The World Bank*. He proposed a "People's World Bank" that would foster "people's globalization" and "grassroots internationalism" as "the next stage of human solidarity," and he wrote, "It gives hypocrisy a bad name when the U.S. government and the institutions it dominates, such as the World Bank and the IMF, go around the world pushing a free-market model on Third World countries when the historical record shows that neither the United States, nor any other wealthy country, used that model."

But like death and taxes, the World Bank is still there, accountable to no one and no government. Dominating Pennsylvania Avenue, its palace of stainless steel, marble, and glass piercing Washington's low skyline, the Bank sits unscathed by critics' attempts to cut off its funding, or by the massive antiglobalization demonstrations in 2000, during the Bank's annual spring meeting, that turned Washington into a police state with all downtown streets closed for three days so global finance ministers and bank officials could meet in peace. This was a year after the huge WTO demonstrations in Seattle, and I sat with my wife in the bright sun on the White House Ellipse, listening to speeches by Ralph Nader, the president of the AFL-CIO, and dozens of angry activists and thought for a moment that I was back at a 1960s antiwar demonstration. We watched the committed young marchers set off from the rally to circle the Bank (outside the closed security perimeter and watched closely by 10,000 DC cops and National Guardsmen). I was especially delighted to see a parade float called "Structural Adjustment," which featured a monstrous, papier-mâché, Godzilla-like villain gobbling up small countries.

SELF had to work with the Bank, in any case, since it provided most of the energy financing for the developing world. The countries where we worked appealed to the International Development Agency (IDA), the arm of the World Bank that provided heavily subsidized, low-interest loans for power "infrastructure" to countries with the lowest per capita income. I wanted the same subsidies for "solar infrastructure," since the people we were serving were the poorest people living in the poorest nations. (By the way, the four arms that make up the World Bank are the International Bank for Reconstruction and Development [IBRD], the IDA, the International Monetary Fund [IMF], and the International Finance Corporation [IFC].)

As Bruce Rich pointed out in *Mortgaging the Earth*, in 1990 the Bank's donors had called upon the Bank to "expand its efforts in end-use energy efficiencies and renewable energy programs." However, the Bank preferred huge and inefficient energy projects because, frankly, it had a lot of money to loan, and in the 90s, energy was the Bank's largest lending "sector." By the middle of that decade, the Bank had itself, in numerous documents, described off-grid solar photovoltaics as the "least-cost method of delivering electricity to dispersed rural houses in the developing world." This assertion grew out of the early work of ESMAP, and it was a victory in itself, but it never found favor among the officials in charge of power generation and "energy infrastructure" at the Bank. When they added up the costs of all the alternative energy programs they approved (some of which were never actually funded), scattered throughout the IBRD, IDA, and IFC, Bank officials figured they were spending billions on solar and renewables, but it was still less than 1 percent of its energy portfolio. In short, a $30 million solar PV project was important to those of us trying to change the world from the ground up, one rural house at a time, but it was just good PR for the Bank.

In 1999 James Wolfensohn, World Bank president, personally hosted a private meeting at the Bank headquarters in Washington for all the major players in the global PV manufacturing and delivery industry. I met him there for the first time, and he promised that when he was in Vietnam the next week, he would fly over one of our solar villages in his helicopter. He did (see Chapter 8). Wolfensohn, a likeable, responsive leader by any measure, was trying to turn the World Bank elephant around and face the solar challenge head-on (which is why he called the high-level meeting on PV), but even he could not budge the intransigent bureaucracy at the end of the day.

These were busy years for SELF. In 1992 I was contacted by the Council for Renewable Energy (CRE) in Kathmandu, Nepal. Somehow this group of urban professionals, businessmen, and electrical engineers who had mostly trained in Russia during the Cold War had gotten hold of SELF's fax number.

Nepal is one of the poorest countries in the world. Its 22 million people live in a largely roadless, mountainous landscape with little contact with the outside world and a per capita income of $200. So many tourists have trekked up the Khumbu Valley to Namche Bazaar and the base of Everest that our popular image of Nepal is a place of Buddhist monks and stupas and prayer flags. In fact, these characterize only this one valley, populated by Sherpa people who came over from Tibet centuries earlier. Sherpas make up less than 2 percent of Nepalese. Most residents are Hindu, and Nepal is the only official Hindu state. (India is 85 percent Hindu, but it is a secular democracy and has no official religion). Democracy came late to this faraway Hindu kingdom, if it ever came at all. The royal family continued to hold great power until all save two or three members were wiped out by the crown prince in a burst of automatic weapons fire during a gathering at the royal palace in 2001.

But all this came later. The real disasters that beset this former paradise were climate change, deforestation, and too many people for the fragile land to sustain. From the plane, as it flew at 20,000 feet, I could see mile-wide slashes of red earth where heavier than normal monsoons had turned once verdant hillsides to slurries of erosion. Meanwhile, the capital was filled with hundreds of international NGOs and their foreign staffs, trying to save the country from itself.

I booked into the Shangri La Hotel in Kathmandu after passing the cremation *ghats* along the river on the way from the airport. Human bodies were being ceremoniously burned, the smoke pouring over the roadway. The hotel's proprietor was a member of CRE, and he offered me a 50 percent discount and thanked me for coming to Nepal. I was soon met, 15 minutes early, by a whole crowd of Nepalese, some in business attire, others in traditional garb, most in colorful wool sweaters and slacks. It seemed I was late for every meeting thereafter by 15 minutes, which I discovered was because Nepal runs its clocks 15 minutes *ahead* of the local time zone for reasons known only to the country's astrologers. (India runs its clocks half an hour out of sync with the world's time zones.)

Tej Gauchan, leader of the delegation, reached out eagerly to shake my hand and welcome me to Nepal. Speaking in perfect English, he proceeded to explain how this professional group came together to spur the government to bring solar power to rural people, but it had no money to do a pilot project. I had, by fax, told them to find three potential sites and pick the best one, and we'd all go visit it when I got there.

Tej said they were also challenging the "hydro guys" in the government, who dominated the debate about renewable energy, even cutting off funds to study the possibilities for solar electrification. It was obvious why the hydro guys had such power. Nepal is mountainous, and lots of water melts down from the Himalayas, so there must be a huge hydro potential. Most of

Kathmandu's electricity is supplied by dams. But there is also a problem: half the population of the country lives in the Terai, part of the northern India plain, which is pancake flat, and half the mountain people live on high ridges, not in dark valleys that have little tillable land and can be flooded. For those on the mountain ridges and upland meadows, basking in sunshine, it was necessary to build small hydroelectric plants, which often had to be located thousands of feet below. These plants required long transmission lines and year-round engineering maintenance because they were often flooded out, silted up, or sitting idle during summer droughts.

"But the water guys don't want to hear this. Nor do the commission agents, who have made great fortunes bringing in donor money to the government for hydro projects, small and large. Hydro is where the money is," said Tej. "We need to use solar. Can you help us do it?" I told him that with all these highly trained electrical engineers committed to rural development working together with SELF, indeed I could.

Tej was one of the most interesting of the many characters I met on my 12-year solar journey. Born and raised in the remote western Himalayan provincial capital of Jomsom, rarely visited by tourists until recently, he was from the kingdom of Mustang. Somehow he made his way to Kathmandu, got himself an education, learned English, got married, and joined the Nepal Air Force, where he learned to fly. When the country decided in the 1970s it needed a national airline, he was sent to United Airline's flight training school in Denver, which trains international commercial airline pilots. When Royal Nepal Airlines bought its first jet, a Boeing 727, it asked Tej and two other Nepalese crewmen to go to Seattle and pick it up. He related over drinks one night at the Rumdoodle Bar in Thamel how close they came to running out of gas on the flight to Kathmandu. "We had ten minutes of fuel left. If we had missed the airport on the first pass, we couldn't have come around again." he laughed. *Many* planes can't make it on the first pass into Kathmandu's terrain-challenged airport.

This highly educated and elite group of CRE volunteers, which included members of the Shahs (a royal family), high officials in the telecommunications ministry (which used PV to power remote mountaintop communication facilities), and other well-placed, middle-aged professionals, seemed out of step with this exotic, very poor, ancient Hindu kingdom. They knew more about solar energy and PV, specifically, than I could ever know. But, like so many NGOs and well-meaning people in the Two Thirds World, they just couldn't figure out how to organize and manage an actual project. And they didn't have any money.

Thanks to the Moriah Fund in Washington and its environmental program officer, Jack Vanderryn, one of the notable leaders in environmental grant making, whom my board member and friend Ollie had known back at USAID, I was assured enough money would be forthcoming to electrify one experimental village in Nepal — its first. I never wanted to visit a remote site, appear as the white Westerner with money, and then have to let them down, but it was difficult to raise funds unless I'd already prepared a real project on the ground. This was always a tricky issue as we put together SELF's projects, and Tej drove the danger home to me by recommending a book.

Before we set off to the former Kingdom of Gorkha, home of the Gurungs (also known as Gurkhas), in west central Nepal to see CRE's selected site, Tej suggested I read *The Lords of Poverty: The Power, Prestige, and Corruption of the International Aid Business* by Graham Hancock, a journalist for the *Times* of London. It was available in every Kathmandu bookshop (Kathmandu has some of the best and most interesting English and foreign language bookshops in the world), so I ran out and got a copy.

A more devastating indictment of the international aid business, what Hancock calls "Development Inc.," has never been published. This 234-page book, as timely today as ever, is worth at least a thousand tons of World Bank project assessment reports, a hundred tons of USAID monitoring and review documents, or a ton of NGO development studies. Hancock attacks the "aid lobby" and the colossal waste inherent in foreign aid budgeted by Western governments and the World Bank. I didn't say "delivered" because, as Hancock points out, very few of the billions of dollars for development assistance ever reach the poor.

"The ugly reality is that most poor people in most poor countries most of the time *never* receive or even make contact with aid in any tangible shape or form .... After the multibillion-dollar 'financial flows' involved have been shaken through the sieve of over-priced and irrelevant goods that must be bought in the donor countries, filtered again in the deep pockets of hundreds of thousands of foreign experts and aid agency staff, skimmed off by dishonest commission agents, and stolen by corrupt Ministers and Presidents, there is really very little left to go around." What is left, he writes, is then used "thoughtlessly, or maliciously, or irresponsibly by those in power — who have no mandate from the poor, who do not consult with them and who are utterly indifferent to their fate. Small wonder, then, that the effects of aid are so often vicious and destructive for the most vulnerable members of human society."

In a country like Nepal, which has been called "over advised and under nourished," he points out that for the price of just one foreign advisor, 30 to 50 Nepalese experts could be hired. Bringing to my attention this powerful indictment of how aid donors operated around the world was Tej's way of subtly pointing out how much foreign "experts" were hated in Nepal. Clearly, the watchword was humility. And listening.

We set off from Kathmandu in two vehicles one cool, foggy, March morning, climbed out of the high valley (which is actually a thousand feet lower than Denver), and headed down the hairpins that descend to the paved east-west highway, which leads along the spectacular Trisuli River valley to Pokhara, Nepal's second city. SELF split the cost of renting one Japanese van and one Landcruiser, plus drivers. Neither CRE nor SELF could ever afford the Landcruisers or Pajeros that were the birthright of every aid agency and government official in every country I visited.

Turning off the paved Privthi Highway, built entirely with foreign aid, which is the only road link between India and Kathmandu, we traversed rough gravel roads through lush valleys. We stopped in small towns for tea and to pay our respects to district officials, then continued along riverbeds and dirt tracks past paddy fields and water buffalos and scattered villages with brightly colored thatched houses until we ran out of road altogether.

"Where are we going?" I kept asking, having been told little about our destination other than it was the finalist of the three villages CRE had reconnoitered for our solar project.

"Up there," said Tej, pointing to what appeared to be a human habitation stuck to the side of a hillside 2,000 feet straight above us. The van became hopelessly stuck in the riverbed.

"But this road doesn't go up there, does it?"

"No, we walk."

We all had packs, and CRE had brought food and supplies and a solar demo kit. As we approached the steep mountainside rising straight out of the fluvial bottom land, 50 or so villagers emerged from the forest, smiling and waving, led by two tall and serious men named Dak and Tek Bahadur Gurung. They were the elders of the small village at the rim of the valley and of Pulimarang, the sonorously named and spectacularly sited village high above. The men were weathered and strong and dignified, talked little, and signaled for the small boys to grab our gear to carry. It was hard to guess their ages, but given that life expectancy in Nepal was 47, I was sure Tek and Dak were much younger than I was.

We walked upward for two hours through the tall forests, which they told me they were carefully preserving. They understood deforestation was destroying Nepal, but they wouldn't allow it here. They also told me, via interpreters, how they had formed a village development committee that would oversee the solar lighting project that CRE had proposed to them. The people had been waiting 20 years for the government to run a power line to their village (we'd left the last power line many miles back) or to put in a hydro station somewhere below, but it never did. They couldn't believe they might actually get electricity, and from the sun!

I asked how long people had lived way up here on this remote mountainside. "About 36 generations," I was told. At five generations to a century, that would be about 700 years. And life hadn't changed much from what it must have been like in the 14ᵗʰ century — except that there were more Gurungs.

We emerged into carefully terraced paddies worked by buffalo that appeared to have been crossbred with mountain goats. It was twilight, and I soon saw just how dark a village without power can be. Small lanterns were illuminating the large, two-story, stuccoed and thatched houses, and people came out on their polished stone verandahs to watch the procession. My knapsack was somewhere in the throng of excited small boys. We navigated the narrow pathways by flashlight, were taken to the second floor of an enormous two-story house facing on a small village square, and were bedded down by our flashlight-wielding hosts, who supplied us with piles of thick wool blankets.

"This house belongs to the Colonel," said Tej.

"Who's the Colonel?" I asked.

"You'll meet him later."

Some of the older CRE volunteers said they were having trouble breathing in the mountain altitude, which I couldn't understand. Pulimarang, like Kathmandu, was only 4,000 feet above sea level. "Wait until you see Annapurna in the morning," said Tej.

In the morning I headed for the "living" village toilet, which was so named because it was filled with millions of wriggling maggoty worms that consumed whatever arrived from above. I emerged in the dawn light, stood on a stone wall that enclosed the path to the flimsy outhouse, looked across the cloud-filled valley, and there was the great Annapurna massif itself, gleaming in the morning sun. It appeared to be only slightly above the elevation of Pulimarang, but at 8,000 meters its summit was actually *four vertical miles* higher than where I was standing. Such is the scale of these mountains that they are beyond sensory comprehension.

I was given the full tour of all 100 houses and farms perched on the green uplands of sun-dappled "Puli" and imagined this was how author James Hilton pictured his Shangri-La in *Lost Horizon*. I saw an unspoiled agrarian community with no modern amenities, where the women all wore colorful embroidery and the children smiled and giggled as if they'd never seen a white man before. They hadn't. "Do they get many visitors?" I asked, knowing the international popularity of the Annapurna Conservation Region, which we could see in all its awesome grandeur across the valley.

"Puli is not on a trekking route," said Tej. I asked again if any outsiders ever came here.

"Well, an English army officer came here about 15 years ago," I was told. "That's the only white man who has ever come to our village."

The officer had been looking after the affairs of the British Gurkhas who had returned home to retire. These were the villages that had provided thousands of Gurkha fighters for the British Empire, first to defend the Raj, then to be sent to the trenches of WWI, where 15,000 perished, and to the battlefields of WWII. They had last fought in the Falklands, some of the local ex-captains told me, speaking through our many CRE interpreters from Kathmandu, who found these Gurungs nearly as exotic as I did. Tek Bahadur had served in Hong Kong, where the last Gurkha garrison was stationed. Soon there would be no more work for them, and the military remittances, which allowed these villages to prosper, would end. However, retirement payments continued to arrive at the ex-soldiers' bank accounts down in the valley towns. It was these funds, the Village Development Committee informed me, that would allow at least 65 of the 100 families here to purchase SHSs. Several had already bought black-and-white TVs in anticipation.

I asked how they were going to get the solarelectric systems, with their heavy batteries, up the mountain. "Just like you got here. Just like everything gets here. Walk." The teenage boys would be the porters.

A community-wide meeting was held in the village square, and Tek and Dak selected eager volunteers for the Solar Lighting Committee. Everyone assembled, about 200 people, with all the women and children on one side of the square and all the men on the other. We took a photo of the watershed event. "I better deliver on this," I said to myself. Before the meeting broke up, the CRE volunteers and I were asked to line up. The village women came toward us in a procession, each bearing an exquisite lei of brightly colored wildflowers with which they garlanded us until we all were nearly suffocated beneath the offering. Twenty wildflower garlands around the neck are heavy! We were also given the ceremonial red-ochre third eye on the forehead, an intrinsic part of Hindu culture. That evening we were treated to a dance exhibition illuminated by the one bright demonstration solar light that we had brought.

Tek and Dak Bahadur (at left), Tej Gauchan, and the author (third from right) receive garlands from citizens of Pulimarang, Nepal's first solar village.

Back in Kathmandu we got down to work, calculating overhead costs, down payments, interest rates, and revolving fund structure and management, and preparing the technical specifics. These were achingly honest people, willing to match SELF's contribution in kind with their volunteer work. Besides CRE, my other partner was the newly formed Solar Electricity Company (SEC) of Kathmandu, run by Dinesh Shah, Jaganath Shrestha, and Yug Tamrakar — two Tribhuvan University professors and one tough local businessman who was the Siemens Solar distributor supplying PV for the telecommunications market. They had already learned how to make their own light fixtures, inverter ballasts, and charge controllers (see chapter 4) in their little workshop. I advised them on how to improve the "user friendliness" of their electronics, based on SELF's experience in other places, and came to trust their engineering knowledge and capabilities. I negotiated a contract for 65 solar lighting systems for Pulimarang, which would grow in a year or so to 100 using income from the village's revolving fund. Even Gurkha families with British army pensions could only afford an SHS with a three-year loan. SELF would purchase the 35-watt solar modules from Siemens in Singapore and airfreight them in. We decided to use solar batteries from Taiwan. CRE and SEC would work together on the installation and long-term maintenance, some of which would be paid for by SELF, and the rest covered by the solar fund. The Village Development Committee was authorized to open a solar project account at the bank down in Damauli town.

Before leaving Nepal, I took the famous sightseeing flight over Mount Everest to celebrate my 50th birthday. When I got back to the James Hilton suite at the Shangri La Hotel, there was a chocolate cake waiting for me that said "Happy 50th!" in white icing. Who knew it was my birthday? I hadn't told anyone in CRE. I called Patti in Washington, but she hadn't ordered it. I invited Tej over to help me eat it, and we celebrated the launch of the first solar village in Nepal.

Back in Washington I followed the progress of the project via fax machine and an occasional airmail letter. I wired grant funds to local banks as needed. Citibank somehow always managed to get them through, and CRE kept track of them to the rupee. People from CRE and SEC made regular trips to check on the project, often traveling by public bus, which took about

14 hours from Kathmandu one way, and then walking the riverbed track and up the mountain trail. All 65 families who had signed up had their solar systems installed by the end of the year and were making their installment payments on the $350 systems to the village solar committee managed by Tek Bahadur. This was the first credit anyone had ever extended to them.

In the spring of 1994 I received a fax asking me to come to Nepal as the country's prime minister and numerous ministers and dignitaries had agreed to officially inaugurate the Pulimarang project. However, when the date was set, I was unable to attend. Later, during a stopover in London, I was invited to meet "the Colonel," at whose antiquated village dwelling we had stayed when I first went to Pulimarang. Colonel Chhatra Gurung, a veteran commander of UN peacekeeping missions around the world, was the military attaché at the Royal Nepalese Embassy in Kensington Gardens, and he asked me over for high tea. He had been the force behind the scene who had directed CRE to his home village, and he was bursting with pride at the success of the solar project, which now illuminated all 100 families. "The lessons learned in Pulimarang are beginning to light up the lives of people throughout the whole of Nepal," he said.

"Mr. Koirala [the PM] flew in to Pulimarang aboard his big Russian helicopter," he told me. "He brought two cabinet ministers, many officials and many press and TV people." The PM had placed a gilded marble plaque in the village to commemorate the inauguration of Nepal's first solar energy project of its kind. Balancing my tea cup, I examined the photographs and press clippings of the event. I had met the prime minister earlier in Washington and had urged him to support a national program based on the Pulimarang model, and he had done just that. Through the country's network of agriculture banks, a 50 percent subsidy was offered to borrowers purchasing SHSs from any of three qualified companies that were just getting into the business: our partner, SEC; Lotus Energy, run by an intense expatriate American Buddhist; or Wisdom Light Group, owned by a Tibetan refugee. Today all three companies are going gangbusters, selling and installing tens of thousands of SHSs all over the country.

Colonel Gurung, a kindly, thoughtful, dignified man in his mid-60s, had accomplished, with the help of SELF and all the local volunteers, his lifelong goal of bringing electric light to his native village, leapfrogging his people from the 14th century directly into the 20th. Because Nepal had a geostationary communications satellite, as well as solar-powered TV broadcast repeater stations on mountain ridges all over the country, it was possible to pick up TV signals easily in the remotest communities, if there was electricity. Colonel Gurung explained that more than half the families had saved up and bought (or military remittances had bought) 12-volt black-and-white TV sets. One retired Gurkha captain had bought a satellite dish and a *color* TV, becoming a fan of British soccer and the world news from the BBC.

The Colonel explained that solar electricity had gotten a bad name in Nepal because four large French-government projects (based on a centralized approach) had failed miserably, so he thought it might be a risk to propose his village as the guinea pig for SELF's first solar-electrification project in Nepal. Accompanying CRE on its scouting mission, he had trekked up to Puli and asked the village elders himself what they thought. "They had never heard of solar power," he told me, "and I explained the solar systems weren't free. It was difficult persuading them of the benefits of solar power. It was hard for them to believe such a thing was possible. Now I get reports that people come from all the surrounding villages to visit Pulimarang to see the solar lights, and they want them for their villages too."

Over the years, academics and aid consultants came from Germany, Denmark, the UK, and Canada to study Puli as a "shining example." One young German, Petra Schweizer-Ries, lived there for several months and wrote her PhD dissertation on the social impacts of electricity on a remote Nepalese community. I wished it was as easy to raise money to do more projects like Puli as it was to get funding to study them and write reports! But we'd done our job with what we had: a small grant and people like Tej and the Colonel and the professionals at CRE, the kind of people who would be the salvation of the country. This was SELF's model, seeking out — or being sought out by — people like this, who cared about their own people and who only needed a small amount of money and a little bit of encouragement to do more with $1,000 than USAID could do with $100,000. In the end, the entire cost of the project that launched a "solar revolution" in Nepal was equal to the duty-paid price of one imported Toyota Landcruiser.

Petra Schweizer

Dak Bahadur and the author in Pulimarang after the installation of the first 75 solar home lighting systems.

☀ ☀ ☀ ☀

Five years later I returned to Puli with a small delegation that included Dr. Wolfgang Palz, PV pioneer and longtime chief of renewable energy for the European Commission, and Dr. Petra Schweizer-Ries, the sparkling young German from the Fraunhofer Institute who'd made Puli the subject of her PhD thesis. I was welcomed back effusively by Dak and Tek, who took us round the village to show us that every solar system was still working fine. Only a few batteries had been replaced, and SEC had upgraded the electronics where necessary, paid for out of the village solar fund. SELF had topped up the fund for battery replacement. Tek showed me the fund ledger, which recorded each villager posting installment payments until the loan was paid back after 36 or 48 months. The fund had "revolved" enough to purchase another 35 SHSs for the rest of the families. TV antennas sprouted from nearly every house; women told me they could weave on their porches after dark under their bright lights; men said they felt, at last, "part of the city" by having TVs and radios. Nepal TV, which had produced a half-hour documentary on Puli, provided educational, agricultural, literacy, and language programs. Once a week the development committee set up a solar-powered TV and VCR on the Colonel's verandah and showed feature films in the village square. Children studied after dark, which they could not do before. "Having light makes us different from all the other villages," said the man who owned the color TV.

And again the women smothered us in wildflower leis. The village children had strung together marigold garlands, and each of them had to bestow his or her garland on one of us as we made our way back down the trail to our rented Landcruisers. We draped the garlands all over the vehicles, which looked like floats in the Pasadena Rose Parade, and headed back to Kathmandu. Later, Pulimarang got its first telephone line. I was given the number of the one telephone, at the Village Development Committee office, and I thought it would be fun to call Tek or Dak, but I never did because they don't speak English and all I could say in Nepali was *Namaste*.

SELF was now hitting its stride. Money was coming in. Our little staff worked long hours and traveled the globe. Bob Freling and I worked up plans for projects that had been requested in Indonesia, Brazil, and the Solomon Islands. The latter came about when I broke my arm hiking in the Alps, and the Swiss doctor who put my mangled wrist back together at his clinic in Meiringen was interested to learn what I did. When I told him, he invited me in my elbow cast up to his all-solar house featuring photovoltaics, solar heat, solar water heating, and passive solar design. He said, "I believe in solar energy."

Dr. Oberli then told me he was giving up his "spoiled good life" in Switzerland to move with his wife to manage the main hospital in Guadalcanal in the Solomons. "I could help you establish a solar program for the Solomons," he volunteered. I sent Bob to the Solomons, and he set up the Guadalcanal Solar Rural Electrification Agency with the help of Dr. Oberli. Soon, bright lights from the high-peaked hardwood and thatch houses could be seen twinkling along remote island shores in the night.

SELF also launched small pilot projects on the Indonesian island of Java and on Brazil's sandy, windswept, northeast coast. Bob was able to put his linguistic skills to work. I continued to commute to China and Vietnam, with visits to South Africa in between, and considered the idea of setting up a base in India, which the World Bank had asked us to do (see Chapter 6).

At home, looking for partners and money, I visited Americus, Georgia, where Habitat for Humanity was based, close to Plains and former president Jimmy Carter's home. After leaving the presidency, Carter had helped Habitat founder Millard Fuller put the organization on the map, and now it was building tens of thousands of houses around the world (90 percent overseas), which poor people could buy with mortgages underwritten by Habitat. These houses didn't come with electricity. Habitat's president liked the idea of joining forces, so we prepared a joint project for Uganda. I raised $150,000 from the US Department of Energy and sent former DOE PV program director Paul Maycock, from SELF's board, to Uganda to set it up. It was a good pilot project, but it didn't catch on Habitat-wide because their houses cost about $700 to finance — yes, a four-room brick house with a metal roof—

Bob Freling

Solomon Islanders display a 40-watt Solarex module at Sukiki, site of SELF's solar village project.

and our SHS cost close to $400. Although you can't normally finance a "consumer durable" over 15 years as you can a house, we did wrap our solar appliance into the cost of the house, but that only made monthly mortgage payments for the new owner higher than most could afford. So after the DOE grant had been tapped out for 150 houses, future Habitat householders got houses without electricity. At the same time, these widely watched and studied efforts launched a batch of private PV enterprises in Uganda, independent of Habitat for Humanity.

In Africa, we also provided funding to electrify health clinics, schools, and community centers in a remote Masai settlement in Tanzania. Working with the young, educated leadership was a delight, especially knowing they were trying to save their people from the encroachment of "civilization" by using the latest technologies available. They needed TV, radios, mobile phones, and computers to survive. Not long after, Masai warriors could be seen walking around with a spear in one hand and a cell phone in the other.

SELF was satisfying a need in many of us to "make a difference." I discovered that *anyone* can, if they want to. You just have to believe, get moving down the road you have chosen, and don't stop. I have always believed that the moment one definitely commits oneself, the universe moves too.

Because SELF was a nonprofit organization under IRS rules and did not need to make money, we were free to experiment and be as innovative as possible. My mantra became, and still is, the "Five Eyes": *Imagination, Ingenuity, Insight, Inspiration, Intuition.* This drove nearly every left-brained person I worked with nuts. I was operating on 100 percent right brain, and hiring the left brainers when I needed them. And I drove them even crazier with the motto I used long before Nike thought of it: "Just Do It!" (Well, actually, it was "Just fucking do it!", or JFDI.)

SELF was "doing it." And validation of SELF's success came in 1996, in a letter from CRE in Nepal that explained how our "solar seed" at Pulimarang had grown into a national program. "A sacred fire ignited by you is spreading," Tej Gauchan had written. Now it was time to spread that fire wider and faster.

But first, I want to tell you a bit about the actual technology you've been reading about here, as well as the manufacturing process, businesses, and corporations involved. Bear with me. It's not *that* technical.

CHAPTER IV

# Technology Time-Out

This is a chapter for laypeople, so if you are already familiar with the solarelectric industry, I won't feel bad if you skip it. As for the lay reader, I promise it won't be too technical. My own understanding of the science behind the "photovoltaic effect" and the complex manufacturing processes involved in producing solar PV modules is minimal. Besides the technology itself, however, I'll also examine the current state of the $50-billion solar PV manufacturing business worldwide.

Solarelectric technology is not as new as some people think. In fact, it is a half-century old. In 2004 the silicon solar cell celebrated its 50th anniversary. Well, the cells couldn't celebrate, but the people involved in developing the conversion of sunlight to electricity celebrated, or at least those who were still alive to mark the date.

The actual discovery of the photoelectric effect is even older, dating back to the Becquerel family, 18th-century French physicists who first experimented with electrochemistry and created a "voltaic cell" that produced a current in 1839. In the 19th century, numerous pioneers working with photochemistry in their laboratories discovered that light had an effect on certain solid materials, such as selenium, creating a flow of current. One inventor, Charles Fritts, sent his experimental "photoelectric plate" to Werner von Siemens (who Americans refer to as the German Edison, while Germans refer to Edison as the American Siemens) who proclaimed that photoelectricity was to be of great importance. In 1876, Siemens himself reported to the Berlin

Academy of Sciences on light's impact on selenium's electrical conductivity. A hundred years later, Siemens, the global manufacturing giant, pumped millions of dollars into its subsidiary, Siemens Solar, which for most of the 1990s was the world's largest PV company.

Initially, solar photovoltaic cells were called "solar batteries" because the voltaic and galvanic (remember the scientists Volta and Galvani?) chemistry that stores electrical energy in conventional batteries is similar to the solid-state chemistry of the *photo* voltaic effect, in which the photons in light, hitting the right conductive material (originally selenium, then chemically treated silicon), knock electrons out of their atomic orbit, creating an electrical current. Sorry if I'm getting too technical here, but I'm really trying to rectify the language and explain how the inelegant and almost unpronounceable name "photovoltaics" became attached to the exceedingly elegant and seemingly magical technology.

It wasn't until Einstein's work on the composition of the atom in the early 20th century provided an understanding of how subatomic particles such as photons and electrons interacted that the actual physics of the photovoltaic effect were comprehended. After that, Bell Laboratories in New Jersey developed the first actual solar cell. In 1954, three scientists at Bell — Daryl Chapin, Calvin Fuller and Gerald Pearson — chemically configured small strips of silicon that could convert 6 percent of the sunlight that struck them into electric current. In 1954 these "power photo cells" were presented to the public as the "Bell Solar Battery." A very small panel of cells powered a 21-inch Erector-set Ferris wheel. The *New York Times* made it a front-page story, and the solarelectric age was born.

This early history, and the ongoing development of PV, is thoroughly chronicled in a marvelous book called *From Space To Earth: The History of Solar Electricity* by John Perlin. His 19-page chapter "Electrifying the Unelectrified," which examines the international off-grid solar market, including the efforts of SELF and SELCO, led me to realize that I had more than enough material and hard-won experience to expand his sole chapter devoted to solar electrification into a book. At the same time, there was no need to restate

Perlin's concise, scholarly telling of how solar electricity came to be. For a greater understanding of this technology and how it grew, I would refer you to Perlin.

The "Space" part of Perlin's title refers to the first successful use, or "application," of Bell Lab's "solar battery." At $286 per watt, Bell figured it would cost a homeowner about $1.5 million to power a house! Bell's Western Electric Company found a commercial application, using small PV modules to boost telephone signals on remote lines in rural Georgia. But silicon transistors came along, which could amplify voice signals at a much lower cost. Then a real market was found: the US Navy Signal Corps was persuaded to put a tiny solar battery on one of its Vanguard satellites in 1958. The first solar-powered transmitter from space worked perfectly, and

rendition by NASA

between Vanguard and Skylab in 1973, and the hundreds of communications and weather satellites launched in subsequent years, an industry was born and numerous semiconductor companies began turning out silicon solar cells. The cost of cells didn't matter, since there was no alternative power source in space.

Photovoltaics power all orbiting satellites and space stations. Here, the Russian space station MIR (with US space shuttle docked below) spreads its solar arrays to face the sun.

The US Air Force, Army, and Navy were all putting up satellites after the Soviets humbled us in 1957 with their Sputnik, the first human-made device ever rocketed into orbit. "Excelling the Russians" was my high-school yearbook's motto. The Russians quickly matched our scientific prowess in solar cells to power their future satellites when Sputnik's nonchargeable batteries failed. Bell Labs and the many researchers working on solar cells had not kept their discoveries secret. So the solar space race was launched. Years later, when I started SELF, I met John Thornton, one of the Army physicists who had worked on the first solar-powered military communications satellites, and he told me that space-grade solar cells he had made in 1961 were still powering signals from some of those original satellites in 1996, 35 years later. (More amazing still, the original solar cell unveiled by Bell Labs in 1954, now on display in a touring exhibit, was still producing a small current in 2004!)

One of the visionaries who helped the solar industry link space and earth was science fiction writer Sir Arthur C. Clarke (author of *2001: A Space Odyssey*), who in 1945 had published a *non-fiction* article proposing that television signals could be bounced around the world from earth to space and back to earth via solar-powered communications satellites positioned in geostationary orbits. That idea certainly looked like science fiction way back then, since few people had heard of television, solar photovoltaics, satellites, or even rockets powerful enough to launch a satellite into orbit. But Clarke saw all this coming. Years later, in Sir Arthur's office in Colombo, Sri Lanka, where he's lived since the 1960s, he told me how happy he was to see SELF fulfilling his vision of bringing solar-powered televisions to the country's rural villages. He also lamented that he was never able to patent the communications satellite, the idea that made Marshal McLuhan's "global village" a reality. It was a technological breakthrough — made possible by the solar cell — perhaps more important than the Boeing 707 and 747.

When I first met Sir Arthur in 1993, our Sri Lankan operations were in daily contact with me in Washington via satellite telephone networks (later replaced by high-bandwidth fiber optics, also something Sir Arthur predicted). And SELF eventually set up solar-powered–satellite Internet links in remote communities in the Amazon and South Africa. Without solar PV on the satellites in space and powering the uplink transmitters, downlink receivers, and computers on the ground, the isolated residents of the Xixuaú-Xipariná Reserve in Roirama State, Brazil, and the grateful students at Myeka High School in Mapaphethe, KwaZulu-Natal, would never have joined the modern world (see Chapter 9).

We had Sir Arthur to thank, and thank him I did every time I was honored to visit him at his large home in Cinnamon Gardens. I updated him on our latest endeavors worldwide, and he showed me the latest downlinks from NASA, received by the satellite dish on his roof. He showed me on his laptop a satellite-generated, infrared, digital photomap of the world that clearly marked, with visible light, the places where humans had electricity. In huge areas of Asia, Africa, and South America there was only darkness. "See," he would say, rolling his wheelchair away from his computer desk, "you have lots of work to do!"

Indeed, there was work to do on earth as it had been done in the heavens. Many of the manufacturing pioneers who had started companies to serve the government-funded space program and telecommunications industry were interested in "terrestrial applications." They wanted to bring their technology down to earth, to serve humanity, replace fossil fuels, and bring power to people beyond the electric grid.

Scientists and engineers in many countries had become so enchanted with this almost magical source of power that they abandoned the scientific method, gave up their engineering principles, and began dreaming. They dreamed of using this technological breakthrough in ways that would have "societal impact" and that would put the "Sun in the Service of Mankind," which was the theme of the first Solar Summit in 1973. One dreamer, Dr. Harry Tabor, a well-known research scientist, published a paper in 1967 proposing "Power for Remote Areas" based on solar cells.

Another scientist-dreamer — and doer — was Dr. Elliot Berman, an industrial chemist who founded the Solar Power Corporation in 1969 with backing from Esso (Exxon). At the Solar Summit in Paris in 1973, according to John Perlin, the company announced it had "recently commercialized and is currently marketing a solar module ... which will compete with other power sources for earth applications." Berman wanted to provide electrical power to those in need, he told Perlin, especially "those who live in rural areas of developing countries." Berman promoted the idea that lower-cost manufacturing using existing silicon wafer technology was the better strategy, rather than focusing on research to make cells more efficient, which was the approach driven by the space program, where costs didn't matter. On earth, costs mattered. While the space program would pay $100 per watt for highly efficient space-grade solar cells, Berman knew he had to bring the costs down to earth to sell on earth, and he did — to $10 a watt. The first market he found was the US Coast Guard, which had 25,000 navigational buoys that needed their batteries replaced periodically. If the Coast Guard could install batteries that would be recharged by the sun, special trips out to the buoys would no longer be needed.

The men who took Berman's dream to the next step were Dr. Joseph Lindmeyer and Dr. Peter Varadi, who founded Solarex Corporation in 1973, shortly after the Solar Summit. With a handful of space program, military, and telecommunication orders in hand, the former communications satellite scientists built a spectacular "solar breeder" north of Washington, DC, in Frederick, Maryland. The breeder, the result of scientists dreaming again, was a factory covered with solar cells that would produce enough power to produce more solar cells. The slab-roofed Solarex building is still in Frederick and is still churning out cells and modules night and day, but it took far more power than they could self-generate to run all the complex, high-tech, fabrication equipment inside. The giant roof array, visible from the interstate highway, provides backup power through large standby battery banks that keep the production line running during power outages.

Solarex expanded, Dr. Varadi built more factories in Europe and Asia, and the company was soon selling solar modules by the megawatt (mW), now the standard unit of measurement for quantities of solar module production. When they hit 1 mW, they celebrated. (A million — *mega* — watts is the equivalent of 20,000 50-watt solar modules.) These "PV shipments" were in addition to the space-grade solar modules Solarex produced for NASA. A big market were "navaids," or solar-powered navigational buoys and foghorns, which were supplied to the oil companies so they could identify their myriad offshore drilling installations and rigs in the Gulf of Mexico and elsewhere.

Another large market was supplying solar modules to the oil and gas industry for use in "cathodic protection" — stopping corrosion on hundreds of miles of pipelines and thousands of wellheads. These solar modules sent a small electrical charge through a pipe into the ground to neutralize corroding molecules. Chevron, Exxon, Texaco, Amoco, and Arco became big customers for Solarex's corrosion-proofing solar solution.

Joseph Lindmeyer died just after I started SELF, but one day I tracked down Peter Varadi. I wanted to learn more about the early days of solar PV. When he gave me his address on the phone, I realized he lived in the large condo building in Chevy Chase right next door to me, and as we worked out where our apartments were located in our respective condo towers, we discovered we could lean out the window and wave to one another! Peter became a friend and supporter of our work in the developing world, and I regret I did not put him on my board of directors when I started SELCO. Peter had sold Solarex to Amoco (Standard of Indiana) in 1983, and he had a miniature copy of the check encased in a Lucite cube on his mantle. He could retire, but instead he busied himself working out global standards for photovoltaic installations that the World Bank and the investment community could adopt.

During the 1970s, Paul Maycock, the DOE's PV program director, was very busy handing out contracts and grants to the solar PV industry to help take it beyond research and into commercial markets besides the space program. The Solar Photovoltaic Energy Research and Development Act of 1978 had committed $2 billion to this technology under President Carter. The objective of Paul's well-funded program, managed by the Jet Propulsion Laboratory in Pasadena, California, was to bring "per watt" costs down to a dollar per watt. There were two ways to do this: by producing higher volumes of cells with efficient production methods, or by producing more efficient solar cells. A great deal of progress was made on both fronts. Unfortunately, the program's early success was cut short when it was dismantled by President Reagan in 1981, and Paul resigned. His prediction of two-dollar-per-watt solar modules by 1990 was pushed back 15 years to 2005.

What saved the industry, however, was the oil companies. When people would ask me, "Whatever happened to solar energy," and I told them, "The oil companies bought it," they immediately reached for a conspiracy theory. "The oil companies want to suppress it" is the paranoid view I've heard reiterated a thousand times. Solar pioneer Bill Yerkes started a small solar company in 1975 to manufacture solar cells for the new terrestrial markets — navaids and cathodic protection. It was quickly purchased by Atlantic Richfield (ARCO), which launched ARCO Solar. Under the leadership of visionary oilman Robert O. Anderson, who also helped found the Aspen Institute, ARCO Solar was slated to be a $1-billion-a-year company by 2000. Spending millions on R&D and state-of-the-art automated manufacturing, ARCO Solar brought the price down to $8.50 per watt. By the early 1980s it surpassed Exxon's Solar Power Corporation, BP Solar (British Petroleum), and even Total Solar, the French oil giant's PV company. Finally, California-based ARCO Solar passed Solarex in sales, becoming the largest solar PV company in the world for a time.

Without the oil companies' money, which they invested partly because of their experience using PV to prevent pipeline corrosion and partly because of the oil crisis of the 1970s, when investing in any other source of energy seemed like a good idea, PV would have remained a niche product. As Charlie Gay, the former head of the National Renewable Energy Laboratory (NREL) and a former president of ARCO Solar, told one early skeptic — me — "Companies don't invest hundreds of millions of dollars in a technology because they want to destroy it."

While today the oil giants own over half the worldwide PV production facilities, it is Japan — which carefully and deliberately entered the photovoltaic business early on — that comprises the rest of the story. The Japanese pioneered the use of PV for lighthouses along their coast in the early 1970s, and electrical conglomerates like Sharp Corporation got into the business very early, followed by Sanyo, Kyocera, and Mitsubishi. By the 1990s, Japan, which has no oil, was covering thousands of residential rooftops with subsidized solar PV, creating a massive market for this "alternative" energy. Today, Sharp is the world's largest manufacturer of PV, and Japan is the world's leading producer. Sanyo and Kyocera geared up production in the 1990s and purchased several American PV manufacturers and retailers. So oil companies don't entirely dominate the PV business, as it sometimes appears, and readers should know that Exxon and Mobil got out of the solar business altogether in the late 1980s.

ARCO Solar was bought by Siemens, which pumped millions into its worldwide operations in the late 1990s, and Siemens Solar became the best of all the solarelectric manufacturers, making the finest product with the longest warranties and the fairest prices. I for one was pleased to see a real electrical manufacturing firm, second only to GE in size, buy what was at the time the world's largest solar photovoltaics firm. This made sense and produced a happy synergy, to use an overused characterization, since Siemens was in the electric, electrical, and electronic business and was also, unlike the oil companies, a manufacturer of durable goods, from hearing aids to massive power-generating turbines. I became closely acquainted with Siemens Solar president Gernot Oswald, who took a deep interest in the developing-world markets I was working in. He visited remote sites of ours, tramping through swamps and across monkey bridges and into jungles. To my knowledge, no other senior executive in the global solar industry had done that. Gernot loved seeing Siemens Solar modules atop thatched huts in Vietnam, Nepal, and South Africa, changing their inhabitants' lives. He even put a 6-kilowatt system on his own house in Munich. He was equally proud of the 5 mW solar PV roof Siemens constructed atop several large buildings at the Munich Messe (trade fair center) in 1999, the largest array of solar PV in the world at the time. He toured me around the installation with the enthusiasm of a small boy (he was in his late 60s at the time).

It would please the conspiracy buffs to know that in 2001, for reasons that are unclear, Siemens decided to get out of the PV business and sold Siemens Solar to Shell Solar, the world's second-largest oil company. And when British Petroleum bought the American oil giant Amoco, BP also acquired Amoco's subsidiary, Solarex. BP quickly bought out Enron's 49 percent share of Solarex and merged Solarex with BP Solar, which then became the world's largest PV company until surpassed by Sharp.

At the dawn of the new millennium, BP and Shell competed for first place in the world of consumer awareness of solar energy. I only wished we'd had one-thousandth of their combined solar advertising budget when we were promoting solar at the DOE. The ubiquitous Shell ads said, "Wish Upon A Star Or Make A Dream Come True?" and showed the Shell logo above a PV module glinting in the sun. Beneath were the words "The sun holds such bright promise as a clean, renewable energy source." Shell explained that it was working to make the dream of solar energy come true. Its TV ads stated that solar, which started out as research, "may become our biggest business yet." BP reportedly spent $50 million on its new corporate sunburst/sunflower logo promoting "Beyond Petroleum" and was criticized in a campaign targeting BP shareholders by Greenpeace International, which pointed out that BP spent more on advertising its new sunburst logo than it did scaling up PV manufacturing at BP Solar.

Maybe the conspiracy theorists have a point.

In the United States, the early market for photovoltaics, which no one wanted to talk about, was largely in northern California and throughout the western Rocky Mountains, where marijuana growers needed "off-grid" power for their houses and processing operations. No grid, no electric bill, no trace. It was estimated that California's largest cash crop in the late 20th century was marijuana, earning billions of dollars for tens of thousands of small growers scattered throughout the state. Solar retailers like Real Goods, Backwoods Solar, and others sprang up, managed by "hippies" who made a nice living selling solarelectric systems to their counterculture colleagues, who were making even more money growing weed. The executives at ARCO Solar and Solarex never inquired as to where all these shipments of solar modules were going, but lots of mom-and-pop solar dealers were paying top dollar for whatever they could get. Marijuana dealers and PV retailers formed an unwitting alliance that contributed more to bringing solar electricity down to earth than anyone wants to admit.

Solar energy appealed to the counterculture, to the disaffected, to society's rebels, and to those simply wanting independence from "the system." Solar could certainly provide that. A citizens' movement evolved to push the solar agenda politically, culminating in the 1970s in Washington in the work of the Solar Lobby, which published *Blueprint for a Solar America* in 1978. Interestingly and ironically, the disconnect — or chasm, rather — between this group of true believers in solar energy and the scientists, engineers, physicists, managers, executives, and investors engaged in actually manufacturing solar hardware was mystifyingly huge. It remains so to this day. For a thorough examination of the politicization of solar energy, especially as it involved the US utility industry, as well as its perceived ability to address the problem of greenhouse gas emissions, I recommend *Who Owns The Sun: People, Politics and the Struggle for a Solar Economy* by Daniel Berman and John O'Connor.

We are not much closer to a solar economy now than we were 25 years ago, and this same constituency, which filled the Washington Mall on Earth Day 2000 beneath the enormous banner proclaiming "Clean Energy Now," has since pinned its hopes on the hydrogen economy, which most of us may not live to see either. But the vision of new energy lives on, as it should. We *are* running out of oil, which will cause oil prices to soar in a few short years, threatening the entire global economy. I'll take a look at the implications of the end of the age of oil in Chapter 10. Meanwhile, the oil economy greenwashes itself with the dream of the solar economy and hides behind the even more distant promise of the hydrogen economy.

During the 1980s I was back in Colorado, doing other things, so when I returned to Washington in 1987 to work for Greenpeace, I had been out of the loop with regard to solar energy for seven years. After my stint at Greenpeace, during which I learned enough about global warming to know that the world had to kick its fossil-fuel habit, I joined Solarex as a consultant for 18 months, and I researched the other market no one was talking about: household solar electrification in the rural Two Thirds World.

This market was booming. It was well known that half the shipments of solar modules to developing countries were ending up serving off-grid customers in rural areas, but it was also perceived as a problematical market since "these people don't have any money." However, the donor agencies did, and they would pay for solar installations in developing countries, while NGOs and consultants would seek to implement the vision of E.F. Schumacher, British author of the worldwide bestseller *Small Is Beautiful*. Schumacher talked about "intermediate technology," and NGOs promoted "appropriate technology." In the African nation of Mali, solar water-pumping systems were developed as an effective life-saving business, and donor money poured in ... rather, it poured into the pockets of the solar companies and the NGOs and consultants like the Intermediate Technology Group, which became IT

Power. Based in the UK, IT Power, under the longtime, tireless leadership of Bernard McNellis, did more in the early days to promote solar as a solution to real problems in the developing world (i.e., using it for vaccine refrigerators, solar pumping, PV for health clinics and government telecom) than any other single entity. It is still going strong. IT Power pioneered community solar "applications" in China, all over Africa, and in South America, and trained some of the people who later became SELF project managers and SELCO entrepreneurs.

SELCO

Even though they did not understand this market in the way they understood the space, navaid, telecom, and cathodic protection markets, the PV manufacturers nevertheless could not help but notice that off-grid electrification was becoming a major driver of their business. But not one solar company was interested in serving this market directly — it was deemed just too difficult. They were right about that! The closest the solar manufacturers wanted to get to the developing-country market was the telecommunications business. It was far easier to meet with communications ministers and sell them a huge order for solar-powered mountaintop repeaters than to traipse around in the rural areas looking

SELCO solar home system (SHS) kit with all components, including deep-cycle battery, fluorescent lights, charge controller, wiring, and switches.

for customers. Thus, an opportunity and a challenge was waiting, which is the subject of this book.

Okay, you've gotten this far without a really technical discussion about how this stuff works. Actually, I'm not going to tell you how it "works" since I really don't know, but I will tell you how it functions and performs — what PV does, not how it does it.

Here goes. First, it is useful to understand that sunlight hits the earth at a "constant" of 1,350 watts per square meter, and it makes its way through the atmosphere to strike the earth when it is directly overhead ("peak sun") with the energy of 1,000 watts per square meter. Photovoltaics can capture only a fraction of this, depending on the efficiency of the silicon solar cells used. A meter-square solar module, not counting the wasted space between the cells, could effectively capture 150 watts at "peak sun" if it was 15 percent efficient. Most modern crystal-silicon solar cells are between 13 and 15 percent efficient. In the laboratory, some cells have proven to be as much as 30 percent efficient, but no production cells have achieved this. Nonetheless, capturing 15 percent of all this free energy is no small accomplishment.

The amount of power a solar module (remember, a module is one part of what is commonly called a solar panel) puts out is calculated in "watts peak," or Wp (the abbreviation of watt, W, is usually capitalized in memory of Scottish mathematician and engineer James Watt). A 35 Wp module will reach that power output in peak sun.

The next thing you may not need to know, but which I'm going to tell you, is that, curiously, a solar cell of any size or type makes half a volt of *direct current*. Don't ask why; that's just the way it is. Although solar modules are rated in watts, which relate to amperes or amps, the important thing in the off-grid business is to know that it can charge a battery. Most large batteries work on 12 volts. Another fact: It takes roughly 17 volts to charge a 12-volt battery. Thus, a solar module must have 36 cells (36 X half a volt or 0.5) to put out approximately 18 volts. None of this is precise because battery science is a black art, based on arcane electrochemical principles that have been studied for over 150 years. No two battery experts agree on how to best charge a battery, so it is a matter of trial and error.

I mentioned that solar cells produce only direct current. This is handy, since batteries can only store direct current. Unfortunately, all appliances in the United States and elsewhere run on *alternating current* (AC), as do compact fluorescent lightbulbs and tubes. This creates another issue: Fluorescent lights are used because they are much more efficient than incandescent bulbs, which *can* operate on DC. But since fluorescent lights only work on AC, the DC output from the battery, accumulating DC current from the solar module or panel, must be "inverted" in what is called an inverter ballast, a small, complex, electronic device that not only converts DC to AC, but also provides the start-up boost needed by all fluorescent lamps.

## Electricity Flow

How solar power works.

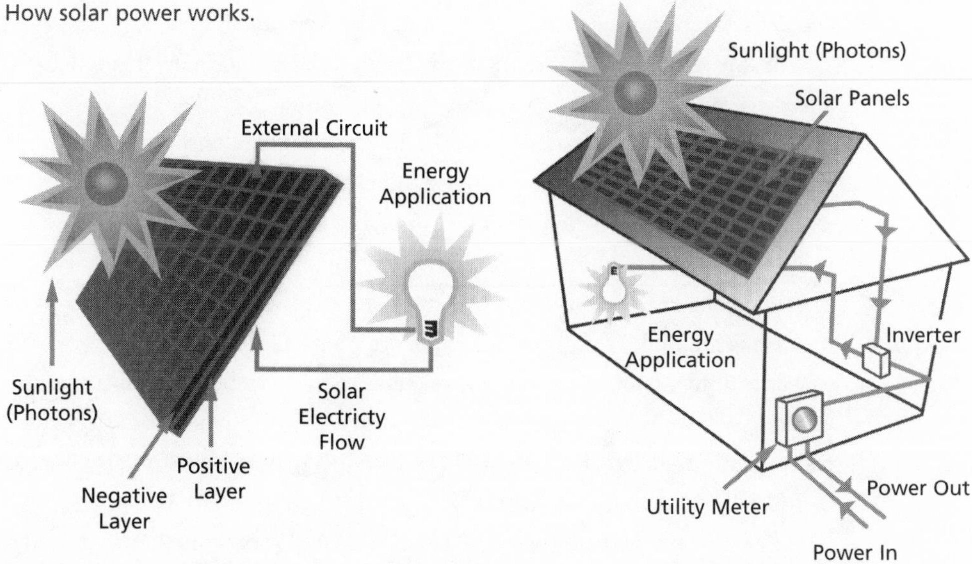

Sunlight (Photons)

Solar Panels

External Circuit

Energy
Application

Inverter

Sunlight
(Photons)

Energy
Application

Solar
Electricty
Flow

Positive
Layer

Power Out

Negative
Layer

Utility Meter

Power In

Spire Corp.

In short, solar modules are simple: they do what they are supposed to do and if properly "encapsulated" — that is, if their cells are sealed between their metal "substrate" and glass — they will last at least 30 years and maybe longer. Batteries, especially the old reliable lead-acid type, are based on a 150-year-old technology. A good workhorse, "deep-cycle" (deep-discharge) battery, like the kind used in golf carts, can last up to seven years. The troublesome part of a solar home system (SHS) is the lights themselves, since the electronic inverter ballasts are prone to failure and early demise. Solar works; it is the parts that fail, known also as "balance of systems" or BOS, that are often the problem.

The other part of any solar installation, big or small, is the charge controller. In the case of small solar-lighting systems using DC power, the controller's job is to make sure the battery does not get overcharged (too much sun) and does not get drained by using too much current. The controller will cut off the system when the battery drops below about 11 volts. Just as in a car, when a battery is drained below 11 volts it stops working. In the case of larger systems, the controller can also be an inverter and convert the DC system power to AC to run common appliances.

In the developing world, most SHSs are DC, and a range of small appliances are available that run on DC.

In the United States and Europe, most solar installations are "grid tied," which means they feed power directly into the utility grid after it has been "conditioned" and inverted to AC to match the sine-wave and cycles of the utility-made power being distributed. (Inverters, by the way, use up as much as 10 percent of the power.) In this case, electric meters run backward when the sun is shining; at night, since there are no batteries in these systems, the power is drawn from the utility power lines. Some 100,000 American households now have rooftop solar systems to augment or replace utility power. If they are stand-alone, they are not connected to the grid and use large banks of specially made batteries instead.

The current cost of solar PV is calculated in "per watt installed," referring to whole systems, by peak watt, or just "per watt," referring to solar modules themselves. Prices have been inching downward as manufacturing capacity increases and production costs go down. You can order one solar module on the Internet for about $4.50 per watt, but you will pay less than $3.00 per watt if you make a volume purchase from the factory.

What else do you need to know? It's useful to understand kilowatt hours (kWh) and how they relate to prices, since conventional electricity is based on the cost per kWh, but solar systems are not (the apples and oranges quandary). At home you probably pay between 7 and 9 cents per kWh, unless you live in the Pacific Northwest, where hydropower still goes for 5 cents per kWh. And you probably use 500 to 1,000 kW per month, multiplied by the kWh price, which gives you your utility bill. However, if you have a business in California, you might be paying 38 cents per kWh at "peak power rates," i.e., midday, when industry, businesses, and air-conditioning are operating.

Today, kWh costs for many solar installations work out to less than 20 cents a kWh. This means solar power is cheaper than conventional commercial power if you have a way to finance the front-end costs of buying and installing it. The utilities build power plants with low-cost, 30-year loans and then charge their customers a flat rate. No household or business can get, or would want, a 30-year loan to purchase a solarelectric system. Herein lies the dilemma of solar. Solar is cheaper than fossil-fuel power *if you can afford it!* Solving this problem on behalf of customers in the developing world has been the biggest challenge faced by companies delivering SHSs, and it has consumed the better part of 14 years of my life. We all had to learn retail banking to sell solar, as future chapters will describe.

How do you calculate kilowatt hours and kWh costs with solar, you might ask? It's quite easy; even I can do it. Before I tell you, however, there is one more thing, which is actually the first thing: *insolation.* This refers to the intensity of the solar radiation striking the earth — i.e., watts per meter squared — at any given place and time of day. Years of research have succeeded in mapping out the solar radiation, or insolation, all over the world. Typically, if cloud cover is not a problem, a solar module in the equatorial regions of the planet will receive six or seven "peak sun hours" a day. During those hours the sun is hitting that area with 1,000 watts of energy per square meter.

SELCO

Compact fluorescent light for DC systems, with self-contained inverter ballast.

In the northern hemisphere, peak sun hours average three and a half to four and a half a day. In South Africa, Saudi Arabia, and western Australia, the sunniest places on earth, peak sun hours can reach eight, since there are few clouds. Weather and cloud cover are considered when calculating peak sun hours. Some of the cloudiest places on earth are right smack on the equator and receive only three peak sun hours per day.

Now that we know the annual average peak sun hours available, we can multiply that number by the size of our solar module. For example, a 50 Wp module in a sunny place with five peak sun hours will produce 250 watt hours each day (note that this is only watt hours; there is not enough produced daily to reach 1,000 watt hours, which is what a kWh is). Thus, six 9-watt lights will use 54 watts per hour (6 X 9), or 54 watt hours (a 9-watt compact fluorescent bulb provides the lumens, or light, equivalent to a 60-watt AC bulb, just like the compact fluorescent bulbs you buy for your home — says so right on the box). If you divide the available 250 watt hours, produced by the solar module and stored in the battery, by 54, you will see that the six lights can operate for 4.6 hours per night. A black-and-white TV connected to the system will use power equivalent to two lights (but measured in amperes), which is why people with solarelectric systems turn off their lights when watching TV. A color TV uses twice the power of a black-and-white set.

It is now possible to compare kWh prices, although it makes little sense to do so for stand-alone PV systems because power is priceless when you don't have any. But let me try: The 50 Wp system described above produces 250 watt hours a day or 7,500 watt hours per month (30 X 250). No one says "watt hours" when they add up to more than 1,000; they become *kilo* watt hours. So this house only gets 7.5 kWh per month! In America we average 500 to 1,000 kWh a month. In other words, approximately 100 times more electric power is consumed in the average American or European home than in a solar-powered house in the Two Thirds World.

So what is the cost of the 7.5 kWh produced by solar PV for the little house in India or Sri Lanka? That depends. If you amortize the $500 SHS over three years, or 36 months, plus 10 percent interest, that's $650 divided by 36, or $18 per month, which works out to $2.40 per kWh. Immediately economists and energy policy types jump in and claim that "solar is too expensive." But they don't realize that the 7.5 kWh of DC current running DC appliances and lights provides almost the same comfort and convenience to a poor family in the developing world as 500 kW delivers to an American family, minus the refrigerator and computers.

If you look at "life cycle" costs, giving the SHS a 15-year average life (less for batteries and electronic components, more for the PV module), and divide 180 months into $650, then monthly costs are $3.60 for 7.5 kWh, or 48 cents per kWh. But forget kWh! These people aren't using and don't need kilowatt hours. They are using *watt hours* quite efficiently and productively. And they get all the watt hours they need for $3.60 a month, amortizing the price plus interest of their SHS over 180 months.

This isn't good enough for most development economists, who will still tell you solar is too expensive. Economists, practitioners of the "dismal science," are people who lie awake at night trying to figure out if what works in the real world might actually work in theory. In the real world, people are quite happy to light up their world for $3.60 month. Even the poor can afford this, especially if each member of a family of six earns $1 a day, which means they bring home $144 a month (working six days a week). They don't pay taxes because they are too poor, and because they are too poor the government or private utilities won't hook them up to the power grid should it come along, since hookups are expensive — usually a third of the cost of an SHS — and it's not worth it to the utility unless the family uses at least 300 kWh a month, which, even at a subsidized five cents a kWh, will cost them $15 a month. So which is cheaper, solar or conventional fossil-fueled grid-delivered power? Never mind the hidden subsidy in all conventional power delivery; if the family had to fund the amortization of the government's power plant over three years, the kWh price would be more like five dollars, not five cents. Instead, the government probably got the power plant for free thanks to a multilateral donor or international investment group, or funded it with tax money on 30-year terms.

This "solar is too expensive" argument came to a halt one day at a World Bank conference when someone noted that the kWh cost of running your wristwatch with battery power works out to $2,000 per kWh! We were told to throw away our quartz watches since they were extremely uneconomical users of outrageously expensive electric power. The point: solar power for much of the Two Thirds World cannot be measured in kWh, any more than apples can be used to measure oranges, or kWh math can be used for a watch battery.

(In the United States, calculating kWh costs of solar is based on more complicated formulas, but it is generally accepted that solar power can be produced in sunny California for 20 cents a kWh, compared to peak power utility charges of 35 cents per kWh for business and industry. This is why the commercial solar PV business in California is booming right now.)

Today solar electricity is a $50-billion-a-year industry worldwide, growing at 30 percent a year. From its humble start, when Solarex celebrated its first 10 mW of module production, annual production worldwide is now 750 mW, approaching its first annual gigawatt. In the United States, the largest producer of cells and modules is Shell Solar (California), which bought Siemens Solar. BP Solar (Maryland) is second, followed by United Solar Systems (Detroit) and Astropower (Delaware). Worldwide, as I've noted, Japan's Sharp is by far the largest manufacturer, followed by Shell Solar, Japan's Kyocera, BP Solar, RWE-Schott (German), Mitsubishi, Isofoton (Spain), Sanyo, and Photowatt (France). The largest PV factory in the Two Thirds World is BP Solar's huge operation in Bangalore (26 mW). In Europe, PV module production in 2003 (rated as "shipments") increased 41 percent over 2002. Only one-seventh of global production took place in the United States, where most of the solar companies are foreign owned. So much for Bell Labs giving America a head start in the silicon solar cell business.

As you are probably relieved to see, we are nearly at the end of this chapter. But there is one more technical item to discuss: production technologies. The companies named above focused on three technologies: monocrystalline, polycrystalline, and amorphous. I'll keep it simple. Monocrystalline, or "single crystal," is the most common, requiring solar-grade silicon to be "grown" or extruded into long ingots in hot furnaces. These ingots are then sliced into wafers from 3 to 5 inches square. This is how all computer chips are made, in case you wondered. It's the same stuff.

Polycrystalline uses waste silicon, with imperfect crystal formations, from the computer industry. It is melted into ingots, which are sliced into paper-thin wafers by hair-thin wire saws. The wafers, either poly or mono, are then "doped" (chemically treated) to function as photovoltaic cells, which convert light hitting their surface into electrons that can be drained off by wire "tabs." As I've pointed out, the average solar module contains 33 to 36 cells (remember, half a volt each, so they can produce the 16 to 18 volts necessary to charge a battery). These cells are strung together and mounted on a plastic-coated steel "substrate," then "tabbed" together in series, covered with a piece of low-iron, tempered glass, and finally "laminated" under high heat to bond with the plastic layer (Dupont Tedlar) beneath the cells. A metal frame is attached, and a solar module is born. Most of this process is automated and often robotized. Making solar cells is an exquisitely refined process and is fascinating to watch, as I've been privileged to do in Bangalore, Detroit, Frederick (Maryland), Hellmond (Netherlands), and Camarillo (California).

The amorphous production technology is also called "thin film." It is a problematical process, and many large production facilities favoring this approach have subsequently been shut down. (Shell wrote off $120 million doing just this in 2001.) The idea is that instead of the time- and labor-consuming process of making silicon wafers, it is possible to deposit silicon and all the other chemicals and bonding agents in a gaseous form directly on metal or glass. Since this can be entirely automated (although using extremely complex and expensive machinery), and since glass is cheap, it was thought this would be the way the world's roofs would be covered with solar. However, amorphous PV is only half as efficient as crystal silicon, so it takes twice as much glass or stainless steel, in modules twice as big as crystalline modules, to produce the same amount of power. And despite the many millions spent by pioneering companies like United Solar, EPV and France's Photowatt, the per-watt price equation has not changed. There are several toxic chemicals involved in the production of all solar PV, but some real bad ones, like gallium arsenide and cadmium telluride, are used in thin-film manufacture. Thus, the workhorse of PV has become the good old monocrystalline and polycrystalline silicon solar cell. The polycrystalline ones are especially appealing to the eye because they are cobalt blue and sparkle in the sun like magic jewels.

I should point out that it takes power to make PV, of course. Nothing is free, not even free power from the sun! It is commonly accepted that it takes power equivalent to one year's output of one 50 Wp PV module to produce that same module. So in a module guaranteed by the manufacturer for 25 years, you really only get 24 years of free electricity. Not bad.

Now I hope you understand why the silicon solar cell ultimately will be to the 21st century what the silicon microchip was to the 20th.

# CHAPTER V

# Born at Pocantico Hills

When historians 1,000 years hence look back at the 20th century, they will see two main things that started then: electricity and oil. Electricity, thanks to Edison, Westinghouse, and Tesla, changed the way we lived. Oil powered the century, replacing wood and providing fuel for transport, which coal could not do.

Before cars appeared on the scene, however, and before electricity became popular, the American oil business was built on the global demand for kerosene, a refined fuel that burned more brightly than whale oil or the vegetable, animal, and mineral oils that humans had been burning in their lamps since they lived in Mesopotamia. The largest fortune the world had ever seen was created by selling refined lamp oil under the manufactured name "kerosene," and John D. Rockefeller became the world's richest man by lighting the oil lamps of the world.

In today's electrically lit world we forget the powerful human need for lighting. It is only brought home during power outages, such as the huge East Coast blackout of August 2003. Everything stops, but worse than that, it's dark! Because the sun is on the other side of the world half of every day, we spend half our lives in the dark. Or we would without lighting. The worldwide demand for lighting is responsible for the third-biggest business (after food and shelter) of modern times — electric-power generation. It's so big we don't even see it, and we take the supply of electricity and electric light for granted.

101

John D. Rockefeller had only one market for his product initially: oil lamps. When kerosene was refined, gasoline was flushed into rivers because there were no cars to use it. Lubricating oil and grease were minor byproducts. Fuel-oil furnaces to heat houses hadn't been invented. Rockefeller might be considered a "robber baron," but he was also the biggest philanthropist America has ever seen, and he did bring kerosene lamps to poor people, since kerosene was cheaper and more plentiful than whale oil, which only the rich could afford. "Let the poor man have his cheap light," he liked to say.

Kerosene was a new fuel, extracted from petroleum coming out of the ground at thousands of wells in Pennsylvania, Ohio, and Indiana, where the global oil business started. Because of impurities from poor refining methods, oil lamps were always blowing up, killing and maiming people, and burning down houses. As Rockefeller consolidated the industry and created a worldwide distribution business that he controlled, he wanted his oil to be safe. He didn't want to kill people. He had standards. So he named his company ... Standard Oil (SO). He said his oil wouldn't blow up because it was "standardized" — carefully refined according to certain safety and production

Century-old cartoon slamming Rockefeller's global control of lamp oil (kerosene) through his Standard Oil Company.

standards. People knew what they were buying. He was not only the world's greatest industrialist, but also the first marketing genius. Safety sells.

At the Museum of the People on Beijing's Tien An Mien Square there is a display showing how the Western oil companies in the late 1800s and early 20$^{th}$ century "exploited" Chinese peasants by giving away oil lamps so that they would buy kerosene. Mounted in a display case were the glass wick-lamps of Standard Oil, Texaco, Gulf, and Shell. Retired oil men I've spoken with remember from their youth the campaign by SO (Esso) "to light the lamps of China." In fact, the Exxon (formerly Esso) corporate magazine is still called *The Lamp*.

When I started SELF, I appealed for funding to the Rockefeller Foundation and the Rockefeller Brothers Fund (RBF), as well as to members of the Rockefeller family, on the basis that hundreds of millions of people around the world were still lighting their homes with smelly, dirty, and dangerous kerosene (not all of it so "standard"). I had a conversation with Larry Rockefeller in his office at the Natural Resources Defense Council in New York one day, and he thanked me for reminding him how much of the original family fortune came from oil (kerosene) for lighting.

Speaking for SELF, I said it was time to replace every 19$^{th}$-century kerosene lamp on earth with a clean, affordable, reliable, 21$^{st}$-century form of energy for lighting, which happened to be solar PV. I made the case that the Rockefeller philanthropies should help end the era of carbon-based lighting, which poured millions of tons of greenhouse gases into the atmosphere annually, and help "bring the poor man and woman their cheap solar light." It worked. A year or so later I was surprised to hear a speech by Rockefeller Foundation president Peter Goldmark, at a renewable energy conference in Washington, in which he borrowed from my own writing and speeches, word for word, phrase by phrase, point by point. In this case, plagiarism was the sincerest form of flattery.

The only difference between our proposed commercial approach and that of J.D.R. a century earlier was — and this is the crux of the matter — that we could not give away lamps and bottle sunlight. Unlike oil, sunlight is free (oil is free, too, if you have the oil lease and the money to get it out of the ground, and then it belongs to the lessee, who can charge for it). If we could lease the sun and sell sunlight, we'd give away the solar cells and the PV modules. But other than that detail, the marketplace for our new lighting product was identical to J.D.R.'s, and it had not changed in over 130 years since he launched Standard Oil back in Cleveland, Ohio, in 1868.

☀ ☀ ☀ ☀

Michael Northrop and Peter Riggs, two program officers at the RBF in New York whom I mentioned in Chapter 3, called me up one day in 1995

Pocantico sponsor Peter Riggs (right), of the Rockfeller Brothers Fund, next to the author at the Vietnam Women's Union solar workshop in Hanoi. (Pham Hanh Sam — see Chapter 8 — is seated at left.)

and invited SELF to organize a small three-day workshop at RBF's Pocantico Conference Center in Tarrytown, New York. I was stunned, honored, and grateful that they thought enough of SELF's activities to make this offer. The conference would be called "SELLING SOLAR: Financing Household Solar Energy In The Developing World."

We put together a short list of 150 invitees. Then the RBF told me that the conference center only had 26 seats, positioned around a big circular table. So we shortened the list, disappointing the uninvited, and came up with a roster of 26 men and women who represented every aspect of the subject at hand: solar electrification in the developing world. I knew all of them personally except for a young, arrogant Wall Street investment banker who someone thought needed to be present but who knew absolutely nothing about the solar business. Each invitee was required to commit three full days to the event. I was reminded of the similar effort to "sell solar" to the American people that Tom Tatum and I organized for the DOE back in 1979 at Gold Lake Ranch in Colorado. This event would turn out to be far more successful and have far-reaching results, at least internationally.

The only thing more boring than sitting through a three-day conference, no matter how interesting the subject matter, is reading about it, so I'll try to spare the reader as best I can. The purpose of the event was summed up in the final white paper issued by RBF: "Three finance-related concerns stand out as being critical if the SHS industry is to grow successfully: First, *purchasers* of solar home systems need to be able to obtain credit from banks or from distributors; second, *manufacturers and distributors* must be able to secure working capital if they are to provide credit to customers; and third, *investors* need credible financing opportunities to move capital toward the solar industry."

To address these issues, we invited leading representatives and principals from the World Bank, the International Finance Corporation, Greenpeace, the Global Environment Facility (GEF), New World Power, Solarex, the United Nations Development Program, the global reinsurance industry, and Wall Street; Paul Maycock of SELF's board, publisher of the industry newsletter *PV News;* several representatives of the philanthropic world, including Bob Crane of the Joyce Mertz-Gilmore Foundation, which was a huge supporter of SELF; K.M. Udupa, deputy manager of Syndicate Bank in India; Dr. Charles Gay, chief of the National Renewable Energy Laboratory (NREL); Richard Hansen of Enersol (see Chapter 2); Rob de Lange, who ran a successful rural solar-delivery company in Indonesia that managed to finance its own customers; Khalid Shams, the number two executive at the famous Grameen Bank of Bangladesh; Deborah McGlauflin, the wonderful facilitator who had been my consultant organizing the Selling Solar conference ... and me. All the people we invited came.

The working papers for the event included a set of case studies SELF had published entitled *Solar Rural Electrification in the Developing World: Dominican Republic, Kenya, Sri Lanka and Zimbabwe*. This little book had had more impact than it probably deserved on international-development policy makers and governments around the world. In it we more or less proved that solar was the least-cost option for millions of unelectrified households, and we showed exactly how, in four countries, thousands of families were improving their lives with electric light and increasing their quality of life with "entertainment communications," World Bankspeak for radios, cassette players, and television. Our report did a lot to demystify the issues regarding electric-power delivery in the Two Thirds World.

The October conference opened with a cocktail party on the exquisite garden terrace of Kykuit, former home of J.D., John Jr., and Nelson Rockefeller, overlooking the Hudson Valley. The Rockefellers' Pocantico Hills estate was named a National Historic Landmark in the 1970s. The RBF leased 86 acres of the estate (including Kykuit) from the National Trust and converted the barns and stables into an elegant conference center with guest rooms, where most of the attendees stayed. The main meeting hall was furnished with an immense round table surrounded by 26 leather chairs. Hanging above it were two 20-foot-wide tapestries, custom-made for Nelson by Pablo Picasso, the only ones he ever did.

I was invited to stay in J.D.R.'s main guest room in Kykuit itself, with views from the second-story windows of the grand entranceway, marble fountain, and huge iron gates through which most of the world's leaders and richest industrialists had passed at one time or another. In this august venue, I thought, we had better get serious.

After the opening dinner hosted by RBF chairman Collin Campbell, I darkened the room and lit a kerosene lamp to demonstrate how two billion people still illuminated their nights. Then I doused the lamp and switched on a compact fluorescent lightbulb connected to a battery, the type SELF provided with its solar home-lighting systems. The room was bathed in cool, bright light. In comparison, the glow from the kerosene lamp could not be called "lighting." However, I pointed out that J.D.R. had called his kerosene "the new light" and the "light of the age." Now, I said, the new light of the age was electric, powered by the sun.

"We're holding this gathering in an appropriate place," I continued. "We want to bring light to the darkness of much of the Third World, and replace kerosene with sun-powered electric light. Well, we're right here at the home of a gentleman who became the richest man in America selling *light* to the world in the form of *kerosene*."

When Standard Oil was broken up by a US court in 1912, I told my audience, over 90 percent of its profits still came from kerosene for lighting, since there was no better alternative. "Now *we're* here at Pocantico to figure out how to give poor people their cheap light .... As wonderful as it might have been in its day, kerosene contributes to global warming, creates health problems for its users, and it can't power a television. Yet half the rural people in the world will still be relying on a carbon-based lighting source from the 19th century as we enter the 21st — unless a new direction is taken by governments, development agencies, energy companies, banks, investors, and, of course, the oil companies for whom kerosene involves much higher profits than photovoltaics."

The timing was right for this conference, I said. It came in the midst of conferences on PV business opportunities and financing decentralized off-grid electrification held in Harare, Manila, Morocco, Nice, and Sun Valley. On the other hand, the World Bank's conference on Financing Sustainable Development, held the week before our Pocantico event, had all but ignored financing sustainable energy. "To the majority of World Bank energy economists and task managers," I said, "300 years worth of coal in the ground is apparently regarded as a form of sustainable energy, and to hell with climate change, greenhouse gases, and global warming. Of course, if we could cogenerate power with all the hot air expended talking about *sustainable development*, we'd displace millions of tons of fossil fuels."

I reminded attendees of the recent cancellation of centralized mega power projects in Nepal and India, of controversial projects going ahead in India and China, and of how AES Corporation recently boasted in the *Washington Post* that the huge coal-fired power plant it was building in Pakistan would bring light to 10 million people. I challenged that statement, pointing out that the transmission and distribution of the power would be cost prohibitive — and that AES wasn't responsible for "T&D." However, the CEO of AES, Roger Sant, was absolutely convinced that solar energy was too expensive, in spite of the examples provided by SELF, Enersol, and other organizations and entrepreneurs.

"I don't think I have to convince too many people in this room that rural people in developing countries will pay for solar electricity in the form of solar home systems," I went on, coming around to the crux of what we needed to discuss at the conference. "But what we do have to work out is how to take, say, $800 million no longer allocated to [the Nepal] dam or a 1,000 mW coal-fired generating plant, and use it to finance an array — to use a good word — an array of decentralized renewables, including the most decentralized of all, solarelectric systems for household power. In Nepal, World Bank task managers have decided to focus on mid-size hydro and diesel generators ... completely ignoring renewables, besides small hydro. This, after the prime minister of Nepal personally inaugurated SELF's village solar PV project last year and called for a national solar program as the only way to reach most Nepalese with electric power for lighting and television. But then, what's a prime minister against a World Bank energy economist!

"If only the world's development economists had listened to E.F. Schumacher when he wrote *Small Is Beautiful: Economics as if People Mattered*. If you've not read it recently, reread it! I doubt we will think of anything here at Pocantico that he hadn't proposed 23 years ago. We need to make small-is-beautiful big enough to count.

"We're here to figure out how a proven, reliable alternative — solar energy conversion technology — can be adopted by 100 million families around the world who have the ability and willingness to pay for it. A third of them can probably pay cash, a third can pay with commercial credit, and a third can afford solar home systems with the help of subsidies.

"We're here to look at financing models and brainstorm new approaches and delivery facilities. Not all the rural Third World is as poor as you might expect. Many rural farmers in the developing world — living *sustainably* — are better off than many of their compatriots in the megacities. But they have no electricity and they are tired of waiting for it. 90 percent of Nepalese, 70 percent of Sri Lankans, 50 percent of Indians, 85 percent of Vietnamese, 75 percent of Indonesians *have no access to electricity* and are denied all the quality-of-life and transformational benefits electricity offers. But they don't want to have to migrate to Kathmandu or Colombo or Delhi so they can see TV, so their kids can read by electric light, so they can quit breathing kerosene fumes. In any case, the lights are going *out* in Kathmandu, Bangalore, Bogota, Manila, and Santo Domingo. Rural people with PV often have lights when city people don't!

"We must consider the issue of subsidies, since most electricity and virtually all rural electricity since the first electricity was generated has been subsidized in one way or another — and rural electrification is still subsidized here in the US. Electricity the world over has always been regarded as a public service.

"And then there is the environment. We can justify a whole lot of bold financial undertakings by the need to develop noncarbon energy sources for the future. As the report last year from the London financial analysts, the Delphi Group, pointed out, 'Climate change presents major long-term risks to the carbon fuel industry. These risks have not yet been adequately discounted by the financial markets. When those discounts occur, clean energy technology will be the beneficiary.'

"But will putting 50 watt-peak solar home systems on 100 million rooftops help the environment? I think it will. Sales of 100 million solar panels will mean photovoltaics is as common as television, and the subsequent price drop will mean Third World cities will start using PV, followed by their industries, then Europeans will use it, and finally even Americans may choose to enter the solar age. And by then huge amounts of carbon will be regularly displaced by clean solar PV."

I ended my remarks with a quote from Dr. Hermann Scheer, founder of Eurosolar and promoter of the Global Photovoltaic Action Plan, a $60-billion proposal to provide a minimum of 10 watts of solar electricity for 1 billion people: "Solar energy, the energy of the people, will offer mankind a prospect of survival that would overcome its spreading fatalism about the future .... Without a radical shift of the world's energy supply systems to non-destructive solar energy sources, without a solar revolution in the wake of the Industrial Revolution, *the Western model of democracy and capitalism is not the perfection of history but its execution*" (my italics). Scheer wrote this in *A Solar Manifesto* (*Sonnen Strategie* in German), published in 1993. (He made an even stronger statement in his 2003 book, *The Solar Economy*, which should be required reading in every business school in America.)

We discussed all these issues, and everyone contributed his or her expertise. The group concurred that the only way solar energy was going to compete with fossil fuels, or bring power and light to poor people in significant numbers, was if we tapped the great river of capital that, in the mid-1990s, was flowing into the "emerging markets" of the developing world in amounts ten times greater than international donor money. New companies had to be formed to replace the nonprofit charity efforts, like SELF, that had hitherto sought to address the problem. Charities simply didn't have enough money to do the job.

The conference was a huge success, for it motivated a dozen people, including me, to try new things. Inspired by this retreat on a beautiful estate amid the great oaks and elms in their fall glory, we all went away on that crisp

October Monday morning trying to figure out how to "sell solar." (RBF later published Pocantico Paper No. 2, *Selling Solar: Financing Household Solar Energy in the Developing World.*)

*Dr. Jeremy Leggett*, an Oxford professor, geophysicist, and expert on climate change (and later the author of *The Carbon War: Dispatches from the End of the Oil Century*) who was then heading up Greenpeace International's Solar Initiative, went back to Britain and launched a new commercial enterprise called Solar Century, now a successful commercial provider of large industrial and residential solar energy systems in the United Kingdom. (Its motto is "If solar can work in England, it can work anywhere.") More on Jeremy later in this chapter.

*Khalid Shams* of Grameen Bank went back to Bangladesh and helped launch the bank's own solar company, Grameen Shakthi, which has subsequently brought electricity to tens of thousands of households in rural areas.

*Richard Hansen*, director of Enersol, the Massachusetts-based nonprofit that had pioneered solar electrification in the Dominican Republic, went home and began to scale up his for-profit venture, SOLUZ, which expanded into Honduras and developed a "solar utility" in the DR, taking up where Enersol left off.

*K.M. Udupa* went back to India and helped SELF's new venture there (see Chapter 6) develop solar loan programs in association with India's largest banks.

*Chaz Feinstein* of the World Bank went back to Washington and pressed the Bank, the IFC, and the GEF, of which he was a senior manager, to finance solar PV whenever and wherever it could. These agencies did just that, for better and for worse, as you'll see anon.

*Dr. Peter Varadi*, my neighbor and the founder of Solarex, whom you met earlier, went back to Chevy Chase and launched a a global accreditation program (PV GAP) to certify rural PV and SHSs so that SHSs would have to meet minimum standards, and international lenders would know what they were financing.

*Dr. Harvey Forest*, chairman and CEO of Solarex, went back to Maryland and focused on international marketing, enabling the company's international dealer networks to better serve, and help finance, their rural customers. I believe he died of a broken heart when BP Solar bought the company, named it "BP Solarex" for about six months, then merged it with BP Solar (its current name — no more Solarex) and based it in Baltimore after forcing Dr. Forest out.

*Dr. Charles Gay*, head of NREL, resigned not long after the Pocantico conference and started a "green business" (Greenstar.com) that used solar to allow local entrepreneurs in remote communities around the world to sell their products, music, and crafts on the brand-new World Wide Web with solar-powered computers and satellite uplinks.

*Brooks Brown*, "a recovering venture capitalist," as he liked to refer to himself, went back to the Environmental Enterprises Assistance Fund in Washington and launched the Solar Development Foundation and Solar Development Capital, which raised tens of millions of dollars to finance dozens of renewable energy companies around the world.

*Christine Eibs-Singer*, who with her colleague, Phil Larocco, had been SELF's entrée to the Rockefeller Foundation through their project-brokering firm LaRocco Associates, went back to New York and co-founded E&Co, a nonprofit investment bank underwritten by the World Bank, the Rockefeller Foundation, and private funds. Phil once told me in his Sixth Avenue office, one floor down from the unapproachable Rockefeller Foundation, "[SELF will] get checks from the Rockefeller Foundation, but you'll never meet anyone there." Right he was, and some very big checks came through for operations in China and India, following project proposals that Phil and Christine vetted. One, for a quarter of a million dollars, was lost in the mail, so the foundation just sent me another. As promised, I never met anyone at the Rockefeller Foundation, and I haven't to this day. It's kind of like the Ford Foundation, whose doors are only open to specially anointed members of a select club, of which I would never be a member since I was always speaking my mind in all the wrong — and many right — places.

*John Kuhns*, a consummate "financial engineer" and founder of New World Power, Inc., the first Wall Street-financed renewable energy company (which specialized in wind power, photovoltaics, and small hydro projects from China to Brazil to Ireland), went back to Connecticut and thought hard about how to scale up the off-grid PV business in those enticing emerging markets. But John wasn't the type to journey out to remote villages to meet and greet potential customers. He had no interest in "sustainable business" or "socially responsible investing." John smelled profits in this vast new energy marketplace. More about John shortly.

I went back to Washington to figure out how SELF could launch a for-profit solar PV company to undertake what we had now proved could be done, but on a significant commercial scale. Deborah McGlauflin had been brainstorming this idea with me for a year or so, and now RBF and the Rockefeller Foundation's representatives were suggesting I set up a private company and raise serious investment capital.

In the spring of 1996 I received a phone call, and not long after there was a knock on the door of SELF's Dupont Circle brownstone office and standing there was S. David Freeman, our solar energy guru from the old DOE days, an energy visionary, and an established major player in the US public utility industry. What the hell was he doing at my door, tipping his rakish cowboy hat?

"Howdy, Neville," he drawled.

I invited him in, and he took a seat in one of the two easy chairs in my office, planting his Tony Lama cowboy boots on the floor. I settled into the other chair after closing the door. I didn't want my staff wandering in.

"I've heard a whole lot about SELF and what you've accomplished," he said, "and I heard you were thinking of starting a company. So am I. I don't think you can do this much longer on a nonprofit basis. It needs to be bigger."

I had been told by people in the utility industry that I should try to contact David, who was then chairman of the New York Power Authority (NYPA), the largest state-owned utility in the nation. As mentioned in Chapter 1, President Carter many years before had put him in charge of the

Tennessee Valley Authority, the largest US-owned electric utility. Subsequently, he'd headed up the municipal utilities in Austin, Texas, and Sacramento, California. At the latter he had instituted revolutionary solar-energy programs, including the construction of a huge PV power plant that partially replaced energy from a nuclear plant the good citizens of California, with David's help, had shut down. In New York he had been trying to institute similar progressive energy conservation and solar programs at NYPA. Friends had told me he was interested in getting NYPA involved with solar energy internationally and offered to schedule a meeting with him, but it never took place.

Now here he was in my office, uninvited and almost unannounced.

"I thought you were very busy running NYPA," I said.

"Governor Pataki fired me. He's a Republican, I'm an outspoken Democrat," David said in his slow, easy, Tennessee accent. "Now I'd like to get back to solar, and maybe I could help you bring solar power to millions of people in developing countries. I'm not too old. I'm 69. I've got another good ten years left. I understand you've got beach heads established all over the world, but now you need Eisenhower and the troops."

Yes, we had "beach heads," which we called pilot projects, and yes, we needed more help, and now here was America's best-known electric-power executive sitting in my office, offering his hand.

This was too good to be true. With David chairing a new company, which I intended to set up and manage, we could raise millions. He knew everyone on God's green earth, from Vice President Al Gore to Maurice Strong, organizer of the Stockholm, Nairobi, and Rio environmental conferences, to Mohammed El Ashry, chief of the GEF, who once worked for him at the Tennessee Valley Authority. He knew the CEOs of half the investment banking firms in New York and London and just about every utility chief in the United States.

"Why do you want to do this?" I humbly inquired.

And he actually replied, "I've got such a huge ego that I can't do anything less than save the world."

He was famous for a wry wit, but this unexpected comment seemed entirely too serious. We agreed to look into the possibilities of setting up a company together. I went to New York for meetings with him, and he jumped on the shuttle a dozen times to come back down to Washington to see me and the new lawyer we had hired to help us put a company together. I began examining the requirements for converting a nonprofit organization to a for-profit enterprise and soon discovered it wasn't possible. Instead, I made plans to resign as SELF's executive director and work full-time for the new entity once we raised some money.

However, there were problems. One of our first disagreements was over the name of the new company. David wanted to call it Sunlight Power. I wanted to call it Solar Electric Light Company, or SELCO, which SELF had used when we established SELCO-India in March 1995. Given this history, David relented and we registered the Solar Electric Light Company as a Delaware corporation in June 1996. It was the last time he would give in on anything.

S. David Freeman had spent his life working for public utilities: he had never worked for, managed, or started a private company. As a powerful chief executive of America's largest public power companies, he was a man to be taken seriously and was accustomed to having his own way. Beneath the cowboy hat, quick wit, friendly smile, and "aw shucks" demeanor was a take-no-prisoners, steel-hearted manager who commanded total allegiance.

"Neville," he told me one day, "I really don't know much about power or finance or management. I'm really just a psychologist." What he was trying to say was he reads people in 30 seconds and either controls them entirely from then on, demanding unyielding loyalty, or dismisses them as nobodies. David was quite surprised, despite a six-month happy honeymoon as we brainstormed what our new company would look like and how, exactly, it would deliver solar power to the world's poor, that he could not order me about as he was used to doing with subordinates all his life. I wasn't a subordinate; we were partners, co-founders and co-owners of a private company.

To me, it was simple. David had dozens of other irons in the fire and was busier traveling the country and meeting people than I could ever imagine being, especially at 69. Thus, he would be SELCO's non-executive chairman, and I would be the CEO.

David had other ideas.

"I'm the boss," he said dryly with a hard edge one day, his disarming drawl failing to charm me this time.

But that wasn't the problem. The problem was that he knew absolutely nothing about the business we were getting into, a business I had already started in India, which was based on six years' experience in 11 developing countries designing and building workable solar finance models, delivery infrastructures, and a suitable technology. And he'd never visited one of these countries to see for himself what it was, exactly, SELF had been doing. On top of that, he had a lot of very bad ideas, and he rarely followed through on things he promised to do. I guessed he was used to having staff follow up.

What he was good at, however, was *vision*. He had a lot of that. He believed quite sincerely that it was our duty to divert as much Western capital as possible to the developing world to finance SHSs for rural people in much the same way America financed power plants over 30 years based on a large customer base of ratepayers. This made sense, to a degree. But we weren't building power plants, and we didn't have ratepayers to finance large capital investments. And we couldn't get 30-year loans! We only had the potential customer base of millions of poor people who nobody wanted to finance, at least not over a 30-year period. Three years was about the maximum loan our customers could expect to get. Despite these differences we struggled on with the model and crafted a draft business plan, and it looked to the world like we were about to be a going concern. David knew there would be no problem raising the start-up capital.

"In America, we abolished the hookup charge in most places. You just signed up for power and you got it, provided you paid your monthly bill on time. This is how rural America was electrified. It should be just that simple for all these poor buggers in these countries," he said. I couldn't agree more, except we'd learned the hard way that it was not "just that simple" to transfer David's power-utility model of doing business to the off-grid, rural PV markets where we worked. It was impossible to raise the necessary front-end capital based on the notion that poor people will pay a "solar utility" to deliver power to them for the rest of their lives. I believed it *was* possible to raise capital on the basis that we would *sell* power systems (and provide service) with sufficient margins to cover the operations, and I believed that we could find third-party lending institutions to make the necessary three-year loans to our customers. In fact, we'd already done that, but without the operating margins.

To investigate how we would raise money, whatever form our business plan took, I arranged to have lunch with David and John Kuhns, whom I had invited to the Pocantico Selling Solar conference. This was a fateful day in ways that would unfold almost immediately and continue to unfold for many years to come.

"I don't think David knows much about private finance," John said afterward. "He's used to going to Salomon Brothers or Goldman Sachs and getting his millions on a handshake, backed up, of course, by the US government and his base of ratepayers. That's not hard. I don't think he understands the business he wants to get into."

John understood how hard it was to market renewable energy in the United States and abroad, as his company, New World Power, had been doing it for some time. Previously, John had formed the largest independent power-generating company in the United States and was operating huge wind farms. He had also financed, privately, California's largest geothermal plant.

"If things don't work out with Dave, call me," John said as he left.

Meanwhile, David and I were spending more and more time with our expensive lawyer in downtown DC, each hour of whose time would buy a poor family in India an entire six-light SHS. Besides crafting corporate documents, the lawyer was noticing, as John had done at lunch, that David and I were not getting along all that well, which did not bode well for a new company. I said, "You're right. This isn't going to work."

David continued to insist, now in public, that he was the boss. He would be the chairman and CEO of the new company, and I'd be the president and would report to him. I began to feel that this otherwise well-intentioned man might be "all hat and no cattle," as they say in Texas.

I said no.

We parted ways soon after. I got the SELCO name back, and David formed a new company with people I had introduced him to, including one of SELF's senior staff, who told me, "The train is leaving the station without you — it's so sad." He also took a financial wizard from Goldman Sachs who I had been intent on hiring. David's fame helped him quickly raise several million dollars, then several million more for Sunlight Power. And just as quickly, leaving no one in particular in charge at his new venture, he took a job as chairman and CEO of the Los Angeles Department of Water and Power (LADWP), the largest municipal utility in the United States. He reportedly fired over 1,000 people, got LADWP into the black, then turned to his main passion, solar energy, and instituted the biggest utility-driven residential solar PV program in the country, much to his credit — and to the credit of his assistant, a remarkable young woman named Angelina Galiteva, who managed the program after they had pushed it through the LA city council and the LADWP bureaucracy.

David moved on to become Governor Gray Davis's "energy czar." He's a hero in California because he kept LADWP — which suffered no blackouts — out of the crooked game with Enron that bankrupted Pacific Gas & Electric and drove California energy prices through the roof. The subsidized solar program he and Angelina set up was a big success, as he instituted in America exactly the kind of solar business he had advocated for the developing world.

However, he left several European investors holding the bag as Sunlight Power, now lacking any serious leadership or direction, soon went under. All that remained was a small, operating, solar utility in Morocco, which a German shareholder had to take over. It was later offered to me, but I passed. This left me to pursue the international solar markets, but I feared that Sunlight Power had poisoned the investor well as news of this fiasco began to spread.

☀ ☀ ☀ ☀

John Kuhns called when he heard that David and I were not going into business together after all.

"I was impressed with the speech you made at Pocantico," he said.

A well-connected investment banker with 25 years' experience in energy markets, John also had an MBA from Harvard Business School. He said he could help me raise money for a solar company. I trusted that he could, and together we founded the Solar Electric Light Company in February 1997.

SELCO, like SELF, took its name from the Edison Electric Light Company and the many other "illuminating companies" formed in the United States at the beginning of the 20th century to generate and sell electric power for the purpose of lighting homes and businesses. Like Rockefeller's Standard Oil, which they soon replaced, the primary business of the Westinghouse and Edison electric light companies was, as their name implied, *lighting*. There was no other use for electricity at the time, just as there was no other use for fossil-fuel oils besides lighting.

I named Bob Freling the executive director of SELF and resigned from the organization. SELCO gave SELF shares amounting to double the appraised value of the SELF projects in India, Vietnam, Sri Lanka, and China. SELF now had a potential future income with which to continue its charitable work, and SELCO acquired the in-country infrastructure in four countries on which to base commercial activities.

John and I, with the help of my assistant Ben Cook, who left SELF to join me at SELCO, worked up an ambitious business plan that we flogged to the big investment houses in New York, Dallas, and Boston. John got the doors opened, but soon they were shutting rapidly behind us as Bear Stearns, Bass Brothers, and other billion-dollar funds politely laughed us back into the street.

"Solar? Hmmm. Third World? You've got to be kidding!"

We tried to explain in our presentations — even to AIG's Emerging Markets Partnership fund in Washington — that these countries offered the biggest untapped power market in the world, and SELCO was going to tap it with solar technology. Solar, we explained patiently, was the only thing that would work where the power grid had failed or did not yet exist. This was big!

I watched a lot of eyes glaze over as I pulled out photos of smiling poor people standing in front of their solar-powered houses. "We're not going to invest in a charity!" one annoyed fund manager said. We explained that although a nonprofit organization had teed up this business opportunity during many years of pioneering development work, SELCO was a totally independent and legitimate for-profit commercial enterprise that expected to make a lot of money for its investors ... one day.

No takers.

Meanwhile, I called Jeremy Leggett, the Greenpeace Solar Initiative rep at Pocantico, who had become a friend. He invited me to attend the Oxford Solar Investment Summit, which he was organizing, co-sponsored by Greenpeace. Dozens of the world's largest insurance and reinsurance companies would convene to figure out how to invest in solar energy instead of the carbon fuel industry, since they had by now all experienced the results of catastrophic climate change attributed to carbon emissions. During the 1990s, the biggest weather-related disasters in history had taken place: Hurricanes Andrew and Mitch, freak storms in Europe, killer heat waves in Chicago and India, and floods in China and Bangladesh. Over the 20th century, a billion-dollar insurance disaster usually happened only once a decade; now, to the insurers' and reinsurers' despair, billion-dollar weather disasters were occurring every year.

The Oxford Investment Summit brought together over 100 UK and European insurance CEOs, bankers, pension fund managers, and solar company executives. It was co-organized by Rolf Gerling, who was also chairman of Germany's insurance giant, the Gerling Group. Rolf Gerling is listed by Forbes as one of the top 300 billionaires in the world. Shortly after the conference, Gerling's Swiss publishing company produced a book, edited by Jeremy, entitled *Climate Change and the Financial Sector: The Emerging Threat and The Solar Solution.* The book was an outgrowth of a 1995 gathering of business leaders (many of whom also attended the Oxford Solar Investment Summit) in Berlin on the eve of the Climate Summit, where 150 nations gathered to negotiate the Framework Convention on Climate Change, which eventually became the Kyoto Protocol.

Gerling and Leggett's book outlined, from the perspective of leaders of the global financial sector, what would happen if global energy demand were met exclusively with fossil fuels. Since insurance and reinsurance companies were the largest single investors in the carbon fuel chain, especially in fossil-fueled power plants, it was paramount that they change their thinking and find new avenues of investment that would help reduce greenhouse gases, which were blamed by nearly every expert for the very real climate disasters threatening the world ... and the insurance business. As far as these insurance and business leaders were concerned, no more studies were needed on global warming and climate change; they already knew what was happening. Now they proposed that their industry be part of the solution instead of part of the problem.

"We were here in Berlin," Gerling wrote, "because most of us, perhaps all of us — I for certain — were worried about global warming and the economic disruption that may accompany it." He pointed out that the key to the Framework Convention on Climate Change was its objective "that governments must take steps to stabilize atmospheric concentrations of greenhouse gases at levels which pose no danger of disrupting the climate in a way that precludes healthy economic activity." He added in the book's foreword, "In facing up to all this, we have to make paradigm shift in our way of life."

A year later, Jeremy Leggett addressed the Oxford gathering, saying, "The core of a clean-energy future — whether based on Shell's energy scenarios or Greenpeace's — requires a huge component of solar energy. Massive solar markets offer the key to a sustainable energy future .... A solar revolution begins to address the population problems, the clean water problem, air quality, urban migration, and nuclear proliferation, among other societal threats."

I delighted in hearing these words. It was "solar revolution" time once again, words I had not heard since our heady days in the Carter administration 15 years earlier. Now, would any of these financial types be interested in investing in the Solar Electric Light Company?

Yes, Rolf Gerling would. Thanks to Jeremy's introduction, Gerling agreed to look at our business plan.

Jeremy also opened doors with another important billionaire. Stephan Schmidheiny, of the famous Swiss industrialist family, was a behind-the-scenes global activist who had financed NGO participation at the Earth Summit in Rio in 1992, and who had also founded the World Business Council on Sustainable Development. Schmidheiny's business manager agreed to look at our plan.

Jeremy's business task force on climate change in Berlin had concluded that PV would be very important in the future, largely because of the markets of the developing world. He had written, "A key global imperative, if governments are ever to reach the agreed objective of the Convention on Climate Change, will be the delivery of sustainable energy for development in the developing world, where two billion people currently have no electricity, and where those who have it will always need more .... The most realistic form of alternative supply, especially in rural settings away from current electricity grids, is solar PV .... Hence, we need to fashion huge global solar PV markets in order to win the endgame in the battle against global warming."

I couldn't have said it better myself. This was the argument that won over Switzerland's richest man and biggest environmental and international development philanthropist. SELCO was there, alone, waiting to get into these rural, off-grid markets in the developing world. We were not looking for philanthropy, but straight commercial investment strictly on the merits of our business.

In the summer of 1997, John and I flew to Europe for meetings in Cologne with Gerling's people at his new investment fund, GAIA Kapital, and in Zurich with Schmidheiny's representative, as well as with a Swiss public investment fund that was raising "green money for a blue planet."

By August SELCO had successfully raised its first $2.5 million of working capital and was ready to go.

We opened a new office in Chevy Chase, Maryland, that I could walk to from home. I didn't want a local car commute on top of my regular 10,000-mile transits to India, Sri Lanka, and Vietnam.

Then I made one of the two or three biggest mistakes of my life: I named two Harvard MBAs to my board of directors, both Wall Street financiers. I was trying to be open-minded instead of listening to gut instincts, which told me I was making a mistake. I hoped these board appointments would reassure our billionaire investors that we were serious about this business. I also felt I needed help because I had no experience managing a company on this scale, but neither of these financiers had ever managed a small business, and they would never really come to understand SELCO's unique mission. We were going to break new ground, learn by doing, and come up with our own formula for creating a profitable, high-growth, but sustainable business, but the non-extractive, non-exploitive business model for the developing world that SELCO was proposing was off the radar of these Harvard MBAs. They may teach sustainability at Harvard now, but they certainly didn't then, and our new venture would never qualify as a typical Harvard Business School case study.

The world of Wall Street and high finance are about as far from the ideals and practices of SELCO and the world it was trying to serve as you can get. I was to discover that most people with Harvard MBAs didn't see the world in the same way as our risk-taking billionaire environmentalist investors or our mission-driven operations managers in the field who were responsible for actually running the company and generating revenues. As Henry Mintzberg, a professor at McGill University, argues in his recent book *Managers not MBAs*, "all MBA graduates should have skulls and crossbones stamped on their foreheads, along with warnings that they are not fit to manage." He criticizes the MBA view of the world as "calculating and analytical" and totally lacking in any understanding of management. (This view was borne out by the corporate corruption scandals involving Enron, Tyco, Worldcom, Adelphi .... They only reflected an extant, arrogant, imperial corporatism that had hijacked global business in the 1990s and would soon give "globalization" a bad name. We didn't want any part of that, even though we were an American-registered, global company financed by some of the wealthiest capitalists on the planet.)

The most positive outcome of all this was a couple of good lunches at the Harvard Club in New York.

I felt better about my convictions when a new investor came along and purchased shares in SELCO. Edward Goldsmith was founding editor of Europe's leading environmental journal, *The Ecologist*, and co-author of *The Case Against the Global Economy*. "Teddy," as he was known, advocated "a turn toward the local," which is exactly what SELCO would be attempting to do as a grass-roots–based, internationally owned, solar service company. Teddy's brother, Sir James Goldsmith, was England's richest man and, right up until his death in 1999, also one of the world's leading anti-globalists.

In the next six years, SELCO walked a tightrope as it sought to become a "new paradigm" mini-multinational corporation beholden to the triple bottom line of social *responsibility*, environmental *sustainability*, and *profitability*. It would take all our *abilities* to achieve this, and I would begin to question whether ethical business practices, socially responsible capitalism, and the principles of sustainability are actually compatible with the demands of finance capitalism.

Meanwhile, I had now moved, for better or worse, from the genteel world of nonprofits and NGOs, where the only bottom line is saving the planet, to the dog-eating, throat-slicing, brutal, and selfish corporate world. I was to discover that I loved business, free enterprise, and working in a market economy, but that I didn't like international finance, so-called free trade, and what modern capitalism and its manipulated and managed markets had become. I read Paul Hawken's *Ecology of Commerce*, which was a turning point for me. David Korten's *The Post Corporate World* opened my eyes even more. I would soon experience firsthand the disparities and contradictions inherent in the global growth economy.

Korten has written, "The global economy is being centrally planned for the primary benefit of the wealthiest one percent of the world's people, a triumph of privatized central planning over markets and democracy." SELCO did not want to be part of this.

Instead, I was about to experience the exhilarating satisfaction of working with young people from the developing world who were organizing a 21st-century enterprise that would not only bring light to hundreds of thousands of people, but would itself be a beacon for anyone seeking a new way for pure market capitalism and fair trade to serve the world. Bringing power to the people would prove to be more fun than I deserved to have.

A few months after I left the nonprofit I'd run for seven years, SELF was nominated to receive the Millennium Award for International Environmental Leadership from Global Green USA and Green Cross Inter-national, chaired by former Soviet prime minister Mikhail Gorbachev. "SELF's commitment to providing a self-sufficient renewable energy path for developing countries is remarkable," read Global Green's announcement letter. "The spirit of the Millennium Awards is to inspire others around the world to follow SELF's cue."

The other recipients were CNN founder Ted Turner and his environmental foundation; Sir John Browne, chairman of BP and a strong proponent of solar; actress Julia Louis-Dreyfus; and Winona LaDuke, who would later be Ralph Nader's vice presidential running mate. SELF was in pretty impressive company! Bob flew to LA, where Mr. Gorbachev presented the award to him at a gala event at the Ritz Carlton. Bob was able to thank Mr. Gorbachev in fluent Russian. My wife, Patricia Forkan, went, as she still served on SELF's board. Patti, along with Paul Maycock (SELF's new chairman), SELF director Larry Hagman ( J.R. Ewing in the TV series "Dallas") and his wife, and Bob and his wife, all shared a table.

I was in India, tramping around the rural areas of Karnataka and organizing SELCO-India in Bangalore, so I missed it.

CHAPTER VI

# Sacred Sun at Your Service

Priyantha Wijesooriya, the executive director of SoLanka, an NGO he and I co-founded in Sri Lanka (see Chapter 7), told me about a fellow student who had studied solar engineering with him in the doctoral program at the University of Massachusetts at Lowell. "He's one of the smartest people I ever met," he said. "But he's very shy, very quiet. I think you should meet him. He's from India."

I faxed Priyantha in Colombo: "Have him call me."

On a very cold day in January 1993, Harish Hande climbed the three flights of stairs to SELF's office on Connecticut Avenue and shyly entered, took a seat, and said nothing. He was wearing only a thin shirt, no undershirt, and he did not even have a jacket with him.

"You must be freezing," I said.

"I'm okay," he said, dismissing my concern and introducing himself.

Maybe he's some sort of yogi who doesn't feel the cold, I thought. After the brief introduction he continued to say nothing. He was very serious and had an aura of great intention and purpose about him, but I wasn't sure exactly what it was.

Harish called me "Mr. Williams," and I mistakenly corrected him, preferring the informal American style of false, friendly familiarity that most foreigners find foreign and strange. He absolutely could not call me by my first name, not someone nearly the age of his father, as such familiarity would be considered extremely disrespectful to elders. I never regarded generation

127

Dr. Harish Hande (right) confers with SELCO's Thomas Pulenkav at a solar house in rural India — note the outdoor porch light.

gaps and age differences as mattering very much, but of course they do. Since I forbade "Mr. Williams," for the next five years he never called me anything, and thus our business relationship was born. His awkward shyness was made even worse by his being unable to address me in any way. He could only stammer out what he was trying to say without prefacing it with "Neville" or "Mr. Williams."

But this didn't stop him, for he wanted very much to work with SELF and find a way to bring solar rural electrification to his home country, India. And he had fire in his eyes, I could see that.

Harish (I'll call *him* by his first name) was studying under the only professor in the United States offering credit courses in solar electricity and solar engineering. He was a graduate of the prestigious Indian Institute of Technology (IIT) at Kharagpur, the Indian university group that is harder to get into than Harvard or Yale. He had recently trained with Richard Hansen in the Dominican Republic, where he had seen "the light" — solar light. He wanted to take Richard's model to India, but Enersol didn't have any money, or any interest in India, so he'd come to see me.

I said that the people at the World Bank's Asia Alternative Energy Unit had been after me to think about starting a solar project in India, and the resident representative of the United Nations Development Program (UNDP) had invited — no, begged — me to come to New Delhi when he heard about our success in Sri Lanka. Foolishly, I had passed up the opportunity, not yet having learned that, next to the prime minister, the most powerful person in many developing countries is the UNDP's "Res Rep," as they're called.

In any case, I had no immediate plans to enter India with a solar project because the size of the place intimidated me, and I wasn't sure how or where to start. India was a bit scary. On my first trip there only two years earlier — a sightseeing excursion from Sri Lanka when I had dismissed the Res Rep — I had been driven right by the spot where, a week earlier, Prime Minister Rajiv Gandhi had been blown to pieces by a female suicide bomber from Sri Lanka's Tamil Tigers. The blood-covered folding chairs were still strewn about the crime scene, which was not even cordoned off. Overlooking the public park where the terrible deed occurred was a stone statue of Indira Gandhi, Rajiv's mother, who had also died violently, gunned down by a Sikh bodyguard. Her carved visage had gazed down from its tall pedestal on her own son's murder. Amazingly, the Indian Air Force didn't bomb the Tiger strongholds near Jaffna in the north of the Island. They could have ended the thug-led Tamil uprising right then and there. As a result, I was beginning to think the Third World was a pretty dangerous place and wondered if India would be stable in the aftermath of Rajiv's assassination.

Harish said not to worry. India was stable. It was a democracy.

I sent Harish Hande to India during a school break to find a good location to begin a pilot solar-electrification program for SELF. For one month he traveled from north to south and east to west. He had lived in Gujarat and Orissa and had family in Karnataka. His father was the manager of India's largest government-owned steel plant, Essar, and the family had moved around a lot. Harish spoke six Indian languages. He and his family were Brahmins, of course, the priestly caste in India.

Harish came back to Washington to see me that spring and announced we could do a project in Karnataka, formerly a "princely state" of the Maharaja of Mysore, and renamed in 1973. It comprises the north end of the Malabar Coast's warm coastal plains, cool highlands amid the Western Ghats, and the southern part of the Deccan Plateau, 4,000 feet above sea level, which is the site of Bangalore, a mid-sized Indian city of 7 million people.

"Why Karnataka?" I inquired.

"Because it is India's richest state and has the worst power cuts, and because I have family in Bangalore and Mangalore," he said matter-of-factly. (Mangalore is a small Indian port city of 1 million inhabitants.)

His assessment was good, but he didn't have any idea how we'd get a project started there. SELF always required local partners, community organizations, and lenders. He didn't know any, nor did I.

That didn't stop us. I went to India shortly thereafter and joined Harish in Mangalore. He introduced me to his remarkable aunt, Hemalata Rao, a self-taught entrepreneur who ran a small cable-TV company out of a shed behind her house. The biggest barrier to her expansion was not a lack of customers, but a lack of customers with electric power. Her cables reached the fringes of the rural areas, where power was intermittent. She understood the need to bring power to the people, which she assured me the perpetually bankrupt State Electricity Board would never do.

Ironically, I arrived during Diwali, the Hindu Festival of Lights, which the city managed to keep lit most evenings. The contrasts of modern India were vividly demonstrated when a band of teenage boys descended upon Mrs. Rao's backyard, where she kept the huge satellite dishes that provided her TV signals, and banged on drums and cymbals, requesting a contribution to the festival. They were all nearly naked and painted dramatically, head to toe, like tigers, with long tails attached to their painted briefs. Their faces were made up to look like tigers, and they were a little frightening at first —

until the American smiled and gave them a donation, and they all burst into big tiger grins. They were happy to pose for photos.

Mrs. Rao took me into her "TV studio," where her stacks of amplifiers, transmitters, and VCRs were the source of a half dozen channels, including her own movie channel. Overhead, in the rafters, a 14-foot-long black whip snake slithered and hovered, looking down at me. "Oh, he lives here," she said. "He eats the rats." Mrs. Rao was a follower of Pharamansa Yogananda, the Indian guru, or teacher, who, following Vivekananda, who came to America in the 1880s, brought Hinduism to the United States in the early part of the 20th century, settling eventually in southern California. (I was something of a devotee myself, having had my life turned around by studying Indian spiritual writers, including Yogananda, author of the perennial classic, *Autobiography of a Yogi*.) Portraits of Yogananda and other Indian saints graced her workshop. She had named her company "Anand Electronics," incorporating this widely used spiritual appellation.

"How big a territory are you licensed to cover?" I asked her.

"We don't have licenses. I just extend my cable service as far as I can reach until I meet the next cable operator's customers," she said. "Whoever gets there first gets the customer. That's how it works."

India, I learned, had some 60,000 independent cable companies like hers. For a formerly closed socialist state with a suffocating bureaucracy, India was inherently a pretty entrepreneurial place. The market economy at this level had not been harmed by Jawaharlal Nehru's and Indira Gandhi's (Nehru's daughter) move to nationalize heavy industry, utilities, banks, and airlines. In fact, it thrived. It occurred to me right then that if we could harness this grass-roots, free-enterprise, business-minded culture to provide solar electricity, we could be very successful.

The Indian electricity problem was bigger than anyone knew or wanted to know. And as soon became apparent, it was especially severe in Karnataka. "There is no more expensive power than no power," Mrs. Gandhi had said when she continued her father's work of electrifying all of India in the spirit of Roosevelt and Lenin, who both put subsidized electrification at the top of their national agendas. But India had no oil and no economic way of transporting coal, and the government's stultifying bureaucracy and socialist policies kept new generating plants from being built or operating efficiently when they did come online.

Thus, India was a well-wired country with no "current," as they referred to electricity, lots of powerless power lines, and a grid infrastructure that might bring one light to the police station in a small community, which

*SELCO-India billboard (called "hoarding" in India) in Mangalore advertising both solar-electric and solar hot-water systems.*

would then be marked on a map in New Delhi as an "electrified village." I was amused by a cartoon in the *Indian Express* which showed a fat, traditionally garbed Indian politician visiting a community where he exclaims, "What! This village has no electricity? But I promised it myself ten years ago!" Politicians here and in Sri Lanka traditionally got elected by promising electricity, but they seldom followed through. So in this, the world's largest democracy, the people would elect new politicians, who would also promise electrification, and again it would not come. Life went on in this great land, and government officials boasted that 90 percent of Indian villages were electrified, while 100 million families, in actual fact, did not have electric power. Government-subsidized, mostly imported kerosene continued to be the lighting method of choice. Harish and I certainly had our work cut out for us here.

We flew from Bangalore to the twin commercial centers of Hubli-Dharwad on the fertile, cool Deccan Plateau, where the British Viceroys had originally planned to put the capital. These were large, dusty, backwater towns of half a million each, serving the agricultural regions surrounding them. In Dharwad, Harish introduced me to K.M. Udupa, a family friend who had taken a liking to Harish and what he was intending to do "for the people of India." Mr. Udupa was chairman of Malabhraba Grameena Bank, a regional rural-development bank with 210 branches, which was owned by Syndicate Bank, one of India's largest government-owned commercial banks. Malabhraba Grameena Bank served over 1,100 villages. Mr. Udupa was in the process of building a modern headquarters with solar-powered backup systems and solar-powered fountains and streetlights. Harish had been hired to install the solar modules. (I introduced Mr. Udupa in Chapter 5, as he was at the Pocantico conference, but this visit in Dharwad was my first meeting with him.)

Over dinner at Mr. Udupa's — I dared not call him by his first name, and never have since, as he is three years older than I — we talked about our respective backgrounds. His wife and daughter hovered in the kitchen, periodically bringing out dishes of savory food. Women never ate with the men.

"I visited the United States once," he told me. "It was 1971. The trip was sponsored by the State Department as part of an agricultural exchange. As a rural banker, I was interested in animal husbandry."

I asked him where he went.

"Ohio," he said.

"I'm from Ohio," I replied. "Where in Ohio?"

"Oh, a small place. You probably never heard of it."

"Try me," I said.

"Chardon, a small village east of Cleveland. Near the Amish communities. We studied their farms."

I smiled and said, "That's where I'm from! I went to Chardon High School. I worked with Amish carpenters in the summer time, had to go to their farms to pick them up since they don't drive."

"Yes, they still use horse and buggies," he said, amazed at this being the case in modern America.

He proceeded to bring out his scrapbook, which contained a photo of Mr. Udupa and my old high school principal Rex Thornburgh, who was on the Rotary Club's committee that had welcomed the Indian delegation. With a billion Indians and some 200 million Americans at the time, I don't need to mention the odds against this happening. We laughed and became fast friends. Mr. Udupa later joined the board of our Indian company.

K.M. Udupa, a short man always nattily dressed in a short-sleeved polyester safari suit, was India's pioneer of biogas digesters. For the uninitiated, these are large, well-like, concrete enclosures that process cow manure and turn it into methane gas that can be burned for cooking and lighting. Tens of millions of them had been sold in India, and millions had been financed by the Syndicate Bank and its rural subsidiaries. They were the ultimate example of "sustainable energy" — provided you had cows.

"We need to do with solar power what we did with gobar-gas," he said, using the Hindi word for cow manure. "There is no other solution to India's rural power situation except solar, and Harish Hande here, I believe, is going to make it happen, with your help."

This was a tall order, but I accepted the challenge of helping Harish fulfill his dream, and Mr. Udupa's dream, of providing solar lighting to every rural family that could afford it. No giveaways or subsidies — this would have to be purely commercial, we knew that. Just like gobar-gas.

"You know, in India, the sun is considered sacred in Hindu spiritual life. Its name is 'surya,' from the Sanskrit. We have worshipped the sun for thousands of years," Mr. Udupa informed me. "We understand the sun's importance in our lives. People already understand solar energy, naturally."

The next time I met Mr. Udupa was in his office at his new posting in Hyderabad, as general manager of Syndicate Bank's main city branch, a big promotion from rural Dharwad. Here he told me the story of India's fifth-largest bank and how it started as a cooperative venture in 1925 in Udupi, a small village near Manipal, Karnataka, a university town a half-hour north of Mangalore, which was where Mr. Udupa was born and grew up. Whereas most large national banks were headquartered in New Delhi, Bombay, Calcutta, or Bangalore, Syndicate Bank's head office was in tiny Manipal, on a hill overlooking the Indian Ocean.

Mr. Udupa gave me two books to read, *The Innovative Banker: T.A. Pai, His Life and Times,* and *The Pais of Manipal* (pronounced "pie" and "pies"). T.M.A. Pai invented rural banking in India, and besides Mahatma Gandhi, a more humble man of the people could not be found in the country. (Gandhi's portrait hangs over the desk of every Syndicate Bank manager.)

T.M.A. Pai and later his son, T.A. Pai, had been Mr. Udupa's mentors from the start back in Manipal. "No man is too small for a bank account," said T.M.A. Pai as he launched his new bank with 8,000 rupees that would offer savings plans to "the little people" to whom banking was a foreign idea. The bank also financed materials for a small weavers' cooperative. Soon farmers, who like most people in the developing world had been accustomed to living hand to mouth, found that their tiny monthly deposits had grown to larger sums than they had ever seen before. The bank was also unique in that it made loans without requiring collateral, relying instead on an individual's "character, credit, and reputation."

The Pais — father, sons, and brothers — continued to carry this family enterprise forward, never forgetting that its success was based on service to rural Indians. "The Indian farmer," said T.A. Pai, "has shown that he has a good business brain. Nobody had to tell him to raise sugarcane, peanuts or cotton. These were profitable cash crops." T.A. Pai was known as "the inno-vative banker." He wanted to help these farmers rise out of poverty, and he saw the key to their prosperity as the village. In the 1950s, under his leader-ship, Syndicate Bank began opening rural branches, which numbered over a thousand when he died in 1981. I was privileged to know his widow, Mrs. Visanthi Pai, who befriended Harish and supported our activities during the following decade, until she too died in 2003.

The Pais' close relationship with Prime Minister Indira Gandhi, who admired the family's enterprising ways and visited Manipal on occasion, did not prevent her from nationalizing the bank in 1975 during "the emergency" in India when she suspended civil liberties and ruled by decree for 19 months in response to political opposition. India's banking sector is still attempting to overcome the legacy of Mrs. Gandhi's policies. Nonetheless, the Pais con-tinued their development crusade, founding large engineering and medical colleges in Manipal. When I was introduced to the Pais in 1995, all they wanted to talk about was bringing solar electricity and hot water to the college dormitories, where nightly power cuts prevented students from studying. We later electrified many campus buildings in Manipal while installing central solar hot-water systems atop the dorms.

Meanwhile, the Syndicate Bank managers, following the lead of Mr. Udupa at his Hyderabad bank post, wanted to electrify their rural bank branches, where power for lighting was rare or nonexistent. We started with the money-counting rooms and vaults, and the bank paid for the PV sys-tems. SELF wasn't too interested in electrifying banks; we wanted to electrify households. However, the banks were the key, and they still are to this day.

I was trotted around to speak at many solar lighting seminars in South India that were sponsored by Syndicate Bank and attended by hundreds of rural branch managers and local government officials. It was usually a big deal to have an American there, and they always made a fuss. I felt a little like a British colonialist, except that I was learning more from these people than they could ever have learned from me. Mr. Udupa introduced me to several powerful chairmen of South India's largest banks, I would be ceremoniously ushered into the great man's enormous chambers, while peons, aides, and numerous supplicants hovered outside, seeking an audience. Mr. Udupa and I would pitch SELF and its mission to bring light to rural people, and the chairman always listened carefully before launching into a lecture on rural development and what to do about it. Mr. Udupa would say, "Ninety percent of the households in India do not have electricity, but the government says ninety percent of villages are electrified." The various chairmen, and the ever-present obsequious managers, would then converse with Mr. Udupa in

rapid-fire Indian English or the local language, and my presence would be forgotten. Having made my opening remarks, I gazed out the windows or studied the colorful wall calendars with their illustrations of

The author and K.M. Udupa, SELCO director, at Syndicate Bank, Hyderabad, training Indian banking officials in solar finance

India's pantheon of gods, took tea when offered, and otherwise remained mute. Afterward, Mr. Udupa would always say of the chairman, "He's with us."

We launched a multi-state training program through Syndicate Bank, SELF, and USAID, which later provided funding support. We educated thousands of bank managers over the years to understand what solar electricity was, how our small household PV systems (SHSs) worked, and how they could be securely financed at the village level. The multi-bank program was later highlighted in a gilded ballroom at the posh Oberoi Hotel in Delhi during a United Nations workshop on "credit mechanisms for renewable energy applications" in 1997. All the blustering bureaucrats had given their monotonous speeches about how much the government and aid agencies were doing to support solar energy (precious little, despite India having a government agency and ministry dedicated to renewable energy) when Harish arrived with the chairman of Syndicate Bank, who gave his own speech about making solar loans at the village level. Manipal's Syndicate Bank had evolved into a huge banking conglomerate, with 1,565 branches nationwide, competing head-on with India's northern-based banking establishment. Mr. Thingalaya, the chairman, arrived in a large white Mercedes, flags flying on the fenders, with two motorcycle police outriders. He upstaged the UN's "alternative energy" bureaucrats at their own conference by actually financing solar power for poor people and not just talking about it. Nothing annoys bureaucrats, anywhere, more than people who actually accomplish things. I smiled as Harish, in his open shirt and sandals, brought the esteemed bank chairman to the podium. This was indeed an unusual alliance. But I'm getting ahead of the story.

Harish and I retreated to my hotel room at the Taj Residency Hotel in Bangalore in October 1994 to figure out how we were going to launch a solar project in India. We pondered and talked and kicked around ideas. Having now seen much more of India, I was no less intimidated by the challenge, despite having met so many amazing, entrepreneurial, and visionary people with the interests of rural India at heart. Whom could we trust as local partners or institutions to bring solar electricity to rural people? We had not yet found an NGO to partner with, and we didn't want to set up our own non-profit group, for that wasn't SELF's approach.

Then it hit me: we could "commercialize" this endeavor right from the start in South India by drawing on the local entrepreneurial spirit and talent *and simply forming a company.* Harish liked the idea immediately. Now what?

I knew India was going to be a difficult place to do whatever it was we decided: form a local NGO, partner with community development organizations, or set up an actual company. The country had been economically and financially closed to the world since independence in 1947, nearly 50 years, becoming self-sufficient in almost all things. Foreign exchange laws were the most restrictive on earth; you couldn't send money in, you couldn't take money out, and the rupee had never been, and still isn't, freely traded. But this was the 1990s, and globalization and the global economy were beckoning, for better or worse.

Rajiv Gandhi had sought to liberalize India and open it to the world economy while promoting and encouraging foreign investment. After his assassination, it was not immediately clear what his successors in the Congress Party would do. But the next prime minister, Narasimha Rao, took the bold step of keeping the country on the path to economic change and liberalization. This meant foreigners could own 51 percent of an Indian company (but not power companies or restricted industries). Foreign companies were not allowed to do business directly or to open branch offices; instead they had to either form a joint venture with an Indian company (as BP Solar did with the Tata family in 1989, forming Tata BP Solar, the largest manufacturer of solar cells and modules in the developing world) or set up a new entity, of which they could own a simple majority of shares.

Harish and I took a three-wheel Bajaj taxi across town to meet with attorneys at one of India's oldest law firms, King & Partridge, something of a holdover from the English barristers who had bequeathed the country a strong foundation in English common law.

I have no gift for languages and can barely understand French, or speak it, despite long study, a year at a French university in Switzerland, and a stint driving a taxi in French-speaking Montreal. But I seem to be able to understand accented English. There are hundreds of versions of English spoken worldwide, but no one speaks faster, with more idioms and odd syntax, than Indians. Anglo-Indians and the upper classes, of course, speak pure Oxfordian English with round vowels that put some BBC announcers to shame. But most Indians speak another English dialect entirely (besides their native tongues of Hindi, Urdu, Kannada, or Malayalam). Communication was everything, and if my tin ear for foreign languages had applied to foreign English, I might as well have packed my bags and gone home. Moreover, Indians often could not understand me except when I spoke very, very slowly, in crisp, clear tones. American accents put them off.

Anyway, my understanding of colloquial Indian English came in handy at the law firm, as the intense "advocate" fired off the steps required for a foreign entity to acquire or form an Indian firm. For a ridiculously low price, King & Partridge agreed to incorporate SELF in India and register us as a foreign-owned firm, requiring special permission from the "permit raj," as India's layered bureaucracy is called.

SELF was a nonprofit, however, and as such couldn't be incorporated under the new investment law or any other laws of India. Besides, we wanted to form a new commercial enterprise that SELF would own and operate, but it had to be separate and independent.

We needed, first, to come up with a name.

As previously noted, the name Solar Electric Light Fund (SELF) was inspired by the Edison Electric Light Company, so named because electric light was its main product, not "power" or "electricity." All electricity in the early days was produced for one purpose only — lighting. (Electricity would have put Rockefeller and Standard Oil out of business, since lighting was their main business too, but then the automobile came along and saved the oil companies.)

It seemed logical that our new for-profit venture should be equally inspired by the early electric pioneers. Solar Electric Light Company sounded about right — until our legal counsel informed us after many a try that he could not register a company with "solar" or "electric" in the title, since these words were already taken. Indian name registration laws are quite different from those in the United States. There was already in Karnataka the "Solar Car Company," so there could not be a "Solar Electric Company." And there was already a company with "Electric" in its name, so we couldn't use that either. (India has the most amazing ways of making the simplest things difficult, if not impossible.)

Since all names usually get reduced to acronyms anyway, we settled on "SELCO" and said we'd call it that. But the company had to state what it did, we were informed, so we came up with SELCO Photovoltaic Electrification Private Limited. That one finally was approved by the licensing authorities, so now we were stuck with the acronym SPEPL!

We decided we would do business as SELCO-India, which became the common name. The registered name was later changed to SELCO Solar Light Pvt. Ltd. since no one could write the actual long name on a check, and no shorthand or abbreviations are allowed on Indian bank checks. At last, on March 14, 1995, SELCO-India was born. King & Partridge had gotten India's first foreign-owned solar energy company approved by the Indian government.

I went back to Washington and began raising money. The first chunks came from E&Co., the private nonprofit investment fund sponsored by the World Bank, the Rockefeller Foundation, and the Rockefeller Brothers Fund. E&Co. took a 5 percent share in the new company. The question soon became: Who else would invest in a commercial venture owned by a nonprofit? As discussed in Chapter 5, we worked that out following the Pocantico meeting when we formed the Solar Electric Light Company (SELCO), which swapped a chunk of its shares for SELF's ownership of SELCO-India. Then the laws changed in India, and SELCO was able to own 100 percent of SELCO-India, which it managed to do after a lengthy appeal to the

President's Foreign Investment Promotion Board in New Delhi. By 1998 we had the first 100 percent foreign-owned solar company in India.

Harish, now finished with his course work in Boston, moved back to India in 1994 and got down to business while he completed his doctoral dissertation, sleeping on his aunt's couch in Mangalore. He also slept on India's express night buses (with reclining airline seats, videos, and music) as he traveled all over South India, setting up the first SELCO solar service centers, the village-based outlets that would become the basis of the business. He trained local youth and hired technicians, managers, and salespeople. He worked tirelessly to establish our new enterprise. On buses or couches, though, I doubted that he ever really slept. His aunt, meanwhile, began producing charge controllers and DC fluorescent lighting fixtures in her workshop.

On my next visit to India, Harish introduced me to Umesh Rai, an illiterate, self-taught, TV repairman from a small town near Mangalore, who quickly took on the mantle of the SELCO mission. If I approved, Harish would hire him as the company's first manager. Here was an Indian Brahmin and graduate of India's most prestigious university sitting down in partnership with an unschooled village appliance repairman to build our little company. How could I say no? Umesh, almost entirely on his own, created SELCO's largest sales and service center, personally made the first several hundred sales — without a phone, an office, or his own transport — and oversaw the installation of each and every SHS in a 20-mile radius. He was a consummate salesman, believing in what he sold, and local people trusted him. These are the kind of human beings that development professionals and foreign aid consultants never meet, but they are the very people who need to be trusted if there is ever to be any success achieved by the international donor agencies. Harish trusted Umesh, and so did I.

(On one of my visits to India years later, Umesh, in his limited English, tried to explain to me how he kept track of exactly how many liters of kerosene we displaced and how he calculated the resulting reduction of $CO_2$ emissions from kerosene lamps. He showed me our total tonnage of "carbon offsets," and we were later able to sell existing and future offsets for a half million dollars to the northwestern US utility PacificCorp and the State of Oregon, becoming the first solar company in the world to find a market for carbon offsets, experimental though it was.)

SELCO-India didn't have money yet for its own transport, so Harish rented jeeps and trucks as necessary, and all employees used their own motor-bikes. One day we jeeped up through some wild hilly country to visit the company's very first SHS installation, in the house of a local landlord who had 25 peasant families working for him, growing coconuts and cashews. His own house was not grand. It already had a biogas cooking and lighting system; the gas cookers worked, but the lights no longer did. I snapped a photo of his wife reaching for the pullcord (before we instituted wall switches as the standard) to illuminate her new living room light. She was so excited she said, "You must send a photo to my sister in Arlington, Virginia." (I was to learn that even in the remotest corners of India, you will *always* encounter someone with a relative in the United States.) The best thing about that day was seeing that the system worked using our own electronics and lights, pow-ered with a PV module made in India by Tata BP Solar. It continued to work and is working to this day, a decade later, with only one change of battery.

Our first family of solar was so happy with their system that they wanted all their sharecroppers to have solar lights as well, so we established the first revolving solar loan fund in India with a grant from SELF. Soon after, room and porch lights could be seen at night glowing from 25 peasant homes scat-tered up and down the verdant hills. The loan fund's operation proved to Mr. Udupa and to local Syndicate Bank managers that people would pay for these systems on installment credit, but collecting the monthly payments proved to be a burden for SELCO. SELCO salesmen began signing up more cus-tomers and taking them to the rural bank branches to help them negotiate three-year loans for $275, the amount after the down payment and front-end cost of wiring the house. This was close to the average annual per capita income of Indians then, so for them, buying an SHS was like us buying a car.

Thanks to Mr. Udupa's relationships with many local bankers, we no longer had to provide the credit or collect on the accounts. The rural bank branches took that on, and other banks followed suit. Five years later Narashima Murthy, chairman of Tungabhadra Gramin Bank, told me "solar lighting systems are just like any other consumer item, except that it has social benefits and it qualifies for anti-poverty, 'priority-sector,' lower-cost loans. We treat solar as a durable good, like a bicycle or tractor or refrigerator or television."

We were now where the American auto industry was when banks first made loans on the horseless carriage, confident that the device would have a useful life at least as long as the life of the loan. We promised the banks in writing that we would buy back any systems they had to repossess, but this rarely happened.

Bangalore was to be the site of the company's head office. It is the capital of Karnataka, a state of 43 million people! It is also India's nicest city, with tree-lined boulevards, vast public parks, grand government buildings, and a pleasant climate year-round. But the pollution caused by diesel trucks and the two-stroke three-wheel Bajaj taxis, which burn a mixture of gas, kerosene, and engine oil, creates an insufferable problem that will only be solved when electricity, natural gas, or hydrogen replaces fossil-fueled transport. The city is also the victim of severe "power cuts," daily scheduled and unscheduled blackouts, which are beyond the ability of the hopelessly inefficient Karnataka Electricity Board to avoid. Bangalore is the "Silicon Valley" of India (and would later become the capital of American-owned "call centers"), but how is this possible without reliable electric power?

It's possible because every business (and every luxury hotel where foreigners stay) has its own diesel generator, thus exacerbating the pollution problem. Nothing could be more hopelessly inefficient, but this resourceful solution works — for now. Meanwhile, rural people, even urban people, remain in the dark. I met the chairman of Karnataka Electricity Board and sought to interest him in solar, but he had other fish to fry, such as keeping his aging generating plants going and his transmission lines maintained. Solar energy was a dream to be deferred, thank you. Anyway, he informed me, 85 percent of Karnataka's village were already electrified, so not to worry, they'd soon get to the rest. In India, bureaucrats believe their own lies, and the higher the official post, the bigger the prevarication. In fact, 50 percent of Karnataka's peasant farmers had no household electric connection, whether they lived in villages or on widely dispersed farms, even if they could see distribution lines along the nearest roadway. Thus, SELCO-India advertised "Electricity at your doorstep."

This was the era of "emerging markets" in the go-go 1990s, when every power utility in the West wanted to build generating plants in India. They spent millions trying to navigate the Indian bureaucracy which, open to investment as it ostensibly was, had no intention of actually allowing any of the proposed plants to get built. I knew this, Harish knew this, but the multinational corporations didn't know this. Huge American firms sent in their naive minions, and signs went up announcing where new power plants would soon arise on expensively acquired or leased vacant land. Mr. Prabhakara, the general secretary of Karnataka (the highest civil office, while the governor is the highest political office), told me that the state's electric crisis would soon be eased, and he itemized the new megawatts of power generation that would soon be coming online thanks to foreign investors. But they never did. When he retired from the Indian civil service, a disillusioned Mr. Prabhakara became one of SELCO's biggest supporters and worked closely with Harish to extend our "solar services" to more rural communities.

We were all watching Enron's plans in Maharashtra, the neighboring state to the north, where Bechtel and Enron were building a 2,000 mW gas-fired power plant called Dhabol. Rebeca Mark, the aggressive Texas executive who was Enron's India project manager, was then a celebrity in India. When Enron's corruption and fraud brought one of America's wealthiest public companies near collapse, US taxpayers lost $700 million, the amount of the Overseas Private Investment Corporation loan to build the thing, while Enron investors lost everything. Ms. Mark, described by *Fortune* magazine as "one of the 50 most powerful women in American business," cashed out of crumbling Enron in 2000, sold her stock, and pocketed $83 million, a pretty nice reward for failing to bring electricity to a single Indian household. All the while, she and her corporate honchos did not understand that no Indian government could raise the electric rates high enough to finance new power plants with foreign investment.

Enron demanded its price, the state said no, and 2,000 mW went undelivered. And this was the other problem: Utility investors were not interested in distributing power in India, only generating it, and at a price set high enough for them to recover their investment over time. Until deregulation, utilities in the United States had to deliver the power they generated; it was part of the deal. Who was going to invest in power transmission and distribution in India? No one.

The best thing about solar energy is that it is already distributed, and can be generated onsite, where it's needed. This little fact escapes most energy planners and power engineers.

We always knew we'd outlast Enron, especially after I heard about Enron's plan to build a 50 mW solar photovoltaic power plant in Rajasthan. (Enron owned half of Peter Varadi's Solarex with Amoco; it also owned a windpower company.) It was the most unworkable idea I'd ever heard of, for a thousand technical reasons I won't go into here except to say, again, who was going to distribute all this power if this central solar plant were built? Certainly not Enron.

But back to our story.

Before Harish could afford an office, or staff, in Bangalore, we needed additional capital. About that time my phone rang in Washington. It was my friend Jon Naar at USAID, offering me money for our project in India. "What could you do with $400,000?" he asked.

"Plenty." I proceeded to draft a budget and write up a proposal. Since USAID never made grants itself, but always operated through "intermediaries" (like all government contracting), we had to get the money from Winrock International, Winthrop Rockefeller's agricultural NGO. USAID funneled a great deal of its international agricultural assistance through Winrock, which was a nonprofit organization like SELF, only many thousand times bigger. After lots of haggling, after several consultants took their cut, and after Winrock took nearly half the money off the top for administration in Little Rock (Arkansas), Arlington (Virginia), and its offices in New Delhi, SELCO Photovoltaic Electrification Private Limited received a check in the mail for US$160,000. Harish was incredulous when he opened the envelope. He never really expected one dollar. This would buy a lot of rupees. The only problem with this "conditional grant" was that it was actually a loan, which over the years SELCO would have to pay back to USAID (not to Winrock, go figure). Winrock publications and PR bulletins later took credit for starting SELCO, even though Winrock had nothing to do with it, and SELF had actually put up more initial capital. This is how the game was played in Washington. As Indira Gandhi said, "There are two kinds of people .... " I won't repeat it again.

Jon Naar later left USAID and joined SELCO as a marketing advisor. He was a leading environmental writer and solar promoter and the author of *This Land Is Your Land* and *Design For A Livable Planet*, two landmark books on planetary stewardship. How he ever got hired at USAID I never knew. He was close to 80 when we took him on part-time, and he traveled tirelessly around rural India taking photos for our marketing brochures (he was also an experienced commercial photographer).

As for the $160,000 conditional grant Jon negotiated for us — never look a gift horse in the mouth, even if it's a loan. The funds allowed Harish to open SELCO's first office in Bangalore, and he hired two experienced "solar salesmen" and business managers who had been involved with solar for several years, working for Tata BP Solar and running their own small solar water-heating shop. M.R. Pai (no relation to the banking family) and Thomas Pulenkav, a Syrian Christian, with Harish in charge, formed the company's managerial triumvirate. Mr. Pai had a PhD in tropical diseases, and Thomas had a degree in business administration. They designed a big sign for the suburban two-story headquarters, and Harish changed couches,

M.R. Pai, SELCO-India manager, speaking at the inauguration of our new Bangalore headquarters in 2001.

moving in with an aunt in Bangalore but still taking the night buses every week to the far reaches of Karnataka, Andrha Pradesh, and Tamil Nadu, where he indefatigably promoted "sun at your service," which became the company's motto.

By 1998, the Washington-based Solar Electric Light Company was capitalized (see Chapter 5) and we could begin a multi-year process of pumping $1.2 million into SELCO-India to fund its growth. Harish and his team — M.R. Pai, Thomas Pulenkav, Suresh Salvagi operating in the north, and other tireless managers — opened solar service centers (SSCs). These were scattered far and wide to serve the unserved and were usually located in small commercial centers, preferably close to a rural bank branch. (There were some 25 of these by 2004, all connected by computers using the company's extremely sophisticated custom management information systems software. The headquarters in Bangalore knew exactly what was going on in any of the rural

SSCs and could track inventory, accounts, customers, personnel, and sales.) Hundreds of SELCO billboards, called "hoardings" in India, and painted wall signs went up: "Make your Home Bright; Use SELCO Solar Light." Money Jon Naar found for us at USAID paid for the production of SELCO television ads for the myriad cable operators to show. Business took off.

Our customers were mostly cash-crop farmers: organic spice growers who shipped their produce to Holland by air; workers on betel nut, sugar, and rubber plantations whose owners sponsored their employees' lighting systems; and growers of cashew nuts, arica (betel) nuts, coffee, coconuts, peanuts, vanilla, cardamom, black pepper, cloves, nutmeg, tumeric, ginger, and tamarind. Many of these farmers refused to use insecticides or chemical fertilizers, and their inherent sense of the rhythms of sustainability and the natural order of things rendered them keen to harness the sun for electricity. They understood.

Nothing was more rewarding for me during the 12 years I spent traveling to "the field" than visiting Indian farmers who had recently purchased an SHS from SELCO. Surprisingly, many of these "poor" farm families lived in substantial houses, or bungalows, with tile roofs, three to six rooms, clean-swept porches, flower gardens, shade trees, and sun-drenched courtyards where they could dry their crops.

The women benefited most, for their kitchens were dark, blackened with soot from cooking fires. We couldn't replace the cooking fuels (wood, coal, briquettes, kerosene), but at least the women could now see what they were cooking! I would be offered flowered leis, glasses of coconut water, sometimes a Coca-Cola or strong coffee. It was rude to refuse, and going from house to house became an exercise in diplomacy. I couldn't imagine Americans opening their houses, or their hearts, in such a way to a company that was showing off the use of its product!

The husband would take me from room to room, proudly turning on the lights and, if they had one, the black-and-white TV. Often he brought out their "retired" lighting systems, which consisted of small tin "bottle lamps," with unprotected wicks that created fire and health hazards beyond imagining.

It was always rewarding when the family noted that their monthly loan payment to the bank was not much higher than their prior monthly expenditure on kerosene and candles — and their SHS could also run a TV, an added benefit for little additional cost.

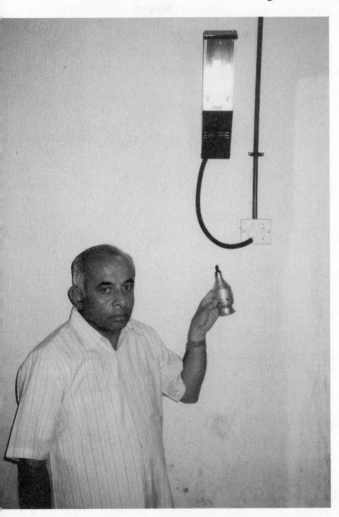

K.M. Udupa, Pocantico attendee and SELCO-India director, with dangerous kerosene bottle lamp displaced by bright new solar light

Sometimes I'd be taken to visit a rich farmer, for whom SELCO had installed a government-subsidized solar water pump powered by 900 kW solar array, 20 times bigger than the standard SELCO household installation. This was a large contraption, mounted in the courtyard on metal poles, that could lift water 100 feet. Farmers often found their solar-powered submersible pump delivered more water than they needed for their fields, so they had excess power which SELCO obligingly — for an additional price — channeled to the house for lighting. It was not uncommon for these farmers to have satellite dishes, color TVs, and lights in as many as a dozen rooms. So much for the "rural poor," I mused.

In the same category was a young "gentleman farmer" who farmed his inherited land for his ailing father, but who made his living trading stocks on the Bombay stock exchange using a solar-powered computer. We drove along narrow back roads to his rambling farmhouse, far from anywhere in central Karnataka, following the telephone poles that brought a single line to the remote farm. He was ten kilometers from the nearest power line. (Telephones operate on a trickle current in the lines themselves; thus Indians commonly have telephones, but no electric power.) Closing the so-called digital divide would become a new business opportunity for SELCO as it later powered Internet kiosks and home PCs with appropriately sized PV systems.

Other customers included urban households and small businesses already connected to the power grid. Some people got fed up with the unremitting power outages and opted for a SELCO light system as a "backup" — and soon found themselves using the SELCO lights even when the utility power was working. One well-off customer in a small town in North Karnataka (he owned a car and a big two-story house) eagerly showed me his sheaf of electric bills, noting how they had decreased as he relied on the solar lights in lieu of grid power. What he saved on his utility bills, he used to pay the bank loan (even well-off customers would not pay cash when bank loans were available — I can't blame them. It was a way to test the system; if it didn't work, they could opt out of paying the bank, which would then come to us for rectification). Why did these households keep the utility connection? Because it was useful to run, even part-time, their refrigerators, which needed only half a day's power to keep food cold. Rural households had no refrigerators, so it wasn't an issue.

There is in Dakshin Kanada (south Karnataka) a remarkable, self-sustaining community that few foreigners ever see. This is Dharmasthala, a mysterious and mystical place run by a powerful religious leader, one Dr. Veerendra Heggade. He represented the 21st generation of "temple managers"

at a highly revered sacred Hindu shrine. (Although the shrine was Hindu, Dr. Heggade was, himself, a Jain, and he'd built a 60-foot statue of the Jain saint Bhagawan Bahubali near the temple.) Mr. Udupa and Harish had been telling me about this legendary figure for years because they wanted his support for a huge solar-electrification project in the communities surrounding "his" temple. I was told he was a "living god."

Besides his ancient Hindu temple, Dr. Heggade also managed an enormous rural development organization called Shree Kshetra Dharmasthala Rural Development Project, or SKDRDP for short. Mr. Udupa was on the board of directors of SKDRDP. Dr. Heggade had a PhD in agricultural and community development, and he had highly experienced international development experts working for him.

In 1999, Dr. Heggade had written in SKDRDP's annual report, "The biggest challenge of the new millennium is achieving development without harming the delicate balance of the planet earth. Struggle for food, shelter and clothing still remain our primary concerns .... Conservation of flora and fauna amidst rapid industrialization is forever a hard balancing act. This necessitates the adoption of a holistic approach to development. Sustainable Rural Development is the only answer to all these questions." This kind of simple wisdom escaped 99 percent of the international rural development "experts" and consultants employed by the World Bank and the UNDP. Dr. Heggade neither wanted nor needed any of their money, for he knew that using it mocked self-sufficiency.

He did, however, cajole the state government into providing him with a subsidy for solar electrification so that 1,500 families could afford to purchase SHSs from SELCO. He organized these beneficiaries into "self-help" groups of five families each, which pooled their money to buy the first system, still expensive for them after the 30 percent government subsidy. The subsidy was paid directly to SELCO, allowing the company to discount its systems by that much and even a little more based on the volume of the project. Development workers trained the householders in how to keep the books on their small loan funds, which they paid in one lump sum, once they had saved it, to SKDRDP, which paid SELCO up front. It took about two years for each of the five families to get their systems. Development occurred, people got electricity, SELCO made a profit, and the government spent less than it would have subsidizing conventional electrification.

"Why couldn't this happen everywhere?" I later asked Mr. Udupa.

"Because there are many vested interests in India who wish to keep people poor," he explained. "So they can be controlled."

Well into SELCO-India's fifth year, I was taken to meet the illustrious Dr. Heggade at his compound deep in the forests of the Western Ghats, about 30 kilometers off the road that winds down to coastal Mangalore from the Deccan Plateau. Harish had explained how they had staged the 1,500-house project during the past year, employing additional technicians and hiring two cooks to feed them in a specially rented facility where they stored the stacks of solar modules and hundreds of batteries that needed to be precharged by a large solar array before delivery to the householders. The region was so vast, with no maps or roads to speak of, that each family who had paid SKDRP for the system had to come in person to pick it up and carry it home. Using some strange homing instinct, our technicians would then follow up with a scheduled visit to install the system in these remote houses scattered through this dense jungle, where the hindrance to solar power was too much shade — so some trees had to go.

I went to Dharmasthala with Harish and Kamal Kapadia, a brilliant young woman I'd convinced SELCO-India to hire after I met her in London while she was studying at Oxford. We arrived in a SELCO jeep and drove under the tall stone gate of Dharmasthala, entering one of the world's most unusual village centers. Here was Dr. Heggade's 1,000-year-old temple to Shiva, with its complement of scantily clad priests. I paid it a visit to be blessed and to honor the small Hindu gods (Shiva, Vishnu, and Ganesh) deep within. Outside, I was also blessed by one of the temple elephants, which lifted its trunk to gently touch my head.

Then we visited the great marbled and columned dining hall where up to 30,000 pilgrims a day were fed by SKDRDP for no charge. I ate there and saw the operation myself, noting copper pots the size of Jacuzzis, filled with lentils (dhal) and rice, in the great kitchen. The tasty repast was served on palm leaves, with everyone seated on the vast marble floor in neat rows.

I was taken to our quarters in the 100-room guest house, owned by Heggade. After a wash and a rest, we walked to Heggade's favorite facility ... his antique car museum. Smack in the middle of these jungles, I was astonished to see an amazing collection of some 30 cars from another era — Rolls, Mercedes, Buicks, and Chevrolets owned by maharajahs. I saw cars like the one I drove in high school, like the ones my father had owned. And they were all kept in running order, although they never went anywhere. Outside the museum, an added attraction for the pilgrims and tourists was a WWII Dakota bomber. Don't ask me why, or how it got there. This is India. The car museum was only surpassed by the enormous art and historical museum down the road, Heggade's very own Smithsonian.

Where was Heggade, I wanted to know? "We will meet him after his daily audience," Harish said. "He is the law of this land, as well as its spiritual leader, and every day over 2,000 people line up to bring him their problems and disputes. He settles them with a few words of advice, which is like the law."

Sure enough, outside his dwelling, across from the temple, was a long line of supplicants and devotees waiting for a few moments with their living god. I was taken into the back of the audience hall to watch the last of the line present themselves to the master. He listened to each one, talked plainly to them in whatever language was necessary, dispensed justice, and cordially sent them on their way. He reminded me immediately of Dr. Ariyaratne of Sri Lanka.

Afterward, he went into the attached house to wash up, and we were ushered into his private quarters. He came out to greet Harish, Kamal, and me. Dr. Heggade settled comfortably into a chair, seemingly unrattled by shaking hands with over 2,000 people that afternoon ... as he did every afternoon. He was large, powerfully built, and had a wide moustache, a wider smile, and big sporty glasses.

Dr. Heggade talked matter-of-factly about his organization and about his other endeavors — his dental college, medical college, and engineering college up on the plateau. I'd noticed these huge modern complexes while we were driving down. He told us how he distrusted government, but got along with the authorities nonetheless, all the while believing only self-reliance would save India, as Gandhi had preached. We talked about the challenges of running a socially responsible business, and he gave me a pamphlet of a lecture he'd given at his business college, entitled "Ethics in Business Management," in which he'd commented how "multi-national and trans-national companies have completely reversed the value systems, ethical standards and social mores" in today's world. He quoted the Rig Veda, one of the Hindu scriptures — "Attempt not to obtain wealth through unjust and condemned ways" — and added, "Never will that wealth be useful which is obtained through violence or oppression. Nonviolent ways of generating wealth without interfering and destroying natural and divine laws is the spirit of living." He claimed his temple was run according to modern management principles of "authority, responsibility, accountability, and transparency." This fellow would never be invited to lecture at Harvard Business School or to address an Enron management forum, I thought.

He presented me with a small silver tabla drum and sent us on our way, but not before saying how proud he was of the SELCO/SKDRDP solar project, which was highlighted in their new annual report. He sent me off particularly to visit his Yoga and Nature Cure Hospital so I could tell people in America about it.

The next day we followed a maze of rough jeep tracks and walked jungle paths to visit a half dozen houses whose owners had purchased SHSs. It was time for flower bouquets again, and coconut juice and strong coffee and smiles and photos, and my shirt dripping wet and sticking to my skin while my Irish complexion got redder and redder in the tropical heat until I looked like a dripping wet tomato. I was humbled by the effusive "thank-yous" from dozens of Indian farmers and their families, who sang the praises of SELCO for all to hear. Coconut juice straight from the husk never tasted better.

People were now buying a "SELCO," as they called our SHSs, validating our brand in a way we'd never imagined. In India, SELCO was becoming the generic noun for solar electricity, as Xerox has become the verb for photo-copying. By 2002, SELCO-India had moved into attractive new offices across the street from the headquarters of Wipro, India's largest software firm. I spoke at the dedication ceremony, which was attended by top Karnataka industrialists and officials, retired functionaries, and notable women, includ-ing SELCO's female managers, who turned out in bright silk saris. I said I believed Dr. Hande was going to become the "Bill Gates of solar energy in India." The rooftop party, beneath a colorful embroidered tent, continued into the night. The three-story headquarters building was entirely outlined in lights, like an extravagant outdoor Christmas decoration in America.

I was very proud that this young team of doers and believers had suc-ceeded at something nearly everyone had said was impossible. They'd done it with almost no help from the "cirque du soleil" (see Chapter 9), the hundreds of international bureaucrats, officials, and consultants who claimed to be spending hundreds of millions of dollars of donor and development agency money on "renewable energy" in developing countries. In India, we saw none of it. Even Winrock was now pressuring Harish to repay its USAID grant, which we offered to do providing Winrock agreed to use the money for more solar projects and not for its own overhead.

The World Bank's and Global Environment Facility's ambitious solar program, announced initially at the Pocantico workshop in 1995, was called the Photovoltaic Market Transformation Initiative or PVMTI. It became our nemesis. Managed by the International Finance Corporation's intransi-gently incompetent and self-important officials, PVMTI nearly killed SELCO and SELCO-India. I could write a depressing case study as long as this book, but I'll try to sum it up in a paragraph:

The $30 million PVMTI project was largely based on what IFC con-tractors and consultants had learned from SELF and SELCO — a great deal from me personally — and a few other solar entrepreneurs. We cooperated for years with World Bank staff while this "initiative" was in preparation, having been told that the program would transform the solar markets and make the dream of "selling solar" that we'd discussed at Pocantico a reality. And we believed it. But it took longer than it took the Allies to launch, fight, and win WWII before the first dollar was lent. By then, highly paid MBAs from McKinsey management consultants and various investment banks had gotten into the game and rewritten the rules to suit their own bottom lines. The leading short-listed company, SELCO, which expected to get a couple of million dollars of the low-cost loan slated for India, was disqualified, and the newly formed Shell Solar India took most of it. So much for a fund formed to help small entrepreneurs bring solar electricity to rural people. We hadn't expected to be competing with the world's second-largest oil company (the largest until the US oil company mergers) for these meager solar funds.

Never mind that Shell had based its operational business plan on ours and sought to use the same banking connections in the same Indian state. Shell had two dozen other Indian states to choose from, but it decided to compete in our backyard and then took most of the pot of development funding we'd been promised over the previous five years. In the end, in 2003, SELCO managed to secure $1 million of low-cost financing from the IFC's PVMTI, plus a "training and marketing" grant of $100,000. By this time they had listened to us, finally, and agreed to make the loan in rupees instead of dollars. We learned the hard way that the last thing the international development banks want to do is development, even if the funds they fear to put at risk are provided by the economically developed nations for international development!

All our solar marketing in India was, nonetheless, attracting attention, including from the Clinton White House, which notified me one day that the president would be travelling to India soon and might like to make a "site visit" to see what SELCO was doing. A Clinton advance team came to SELCO's headquarters in Bangalore, and I provided stacks of information from Washington. Then an invitation arrived addressed to Dr. Harish Hande, requesting his presence at the signing in Agra of the India-US Clean Energy Agreement, which was intended to put $500 million into renewable energy in India. Harish was the only representative from the Indian or US solar industry invited. President Clinton was widely welcomed in India in 2000, although he scrapped Bangalore, and the SELCO stop, for Hyderbad. Harish, however, did attend the formal signing ceremony in the shadow of the Taj Mahal, and we all worried about what he would wear.

No clotheshorse, Harish usually wore sandals or scuffed shoes, no belt, and an open shirt, whatever the occasion. Indians are casual, except for affairs of state, when formality rules. Harish told me afterward that he had taken a seat in the small reception area (only about 100 people were invited) next to Priyanka Chopra, Miss India. He asked her to stick by him and he'd make sure she got to shake hands with Clinton. But he had an ulterior motive — he opined, correctly, that Clinton would come over to meet him if she were there, since no one could take their eyes off the lustrous beauty-contest winner. Harish figured he got in an extra couple of minutes with the president as Clinton gazed at her beauty while listening to Harish talk about solar. Ms. Chopra said nothing. She became famous a week later when she won the Miss World contest.

"So what did you wear?" I teased Harish over the clear, fiber-optic telephone line to India.

"Same thing as the president. Open shirt, casual slacks. Everyone else was in suits and ties." On TV I'd seen Clinton, with daughter Chelsea, at the Taj Mahal, where he'd been prior to the clean energy reception, and, sure enough, he was wearing a red sport shirt and slacks. Afterward, instead of hanging around with Ms. Chopra, Harish noticed a "sort of wallflower, all alone" at the reception and went over and talked to US Secretary of State Madeleine Albright for the rest of the event. Very noble.

SELCO-India would take no Indian government contracts because its management was too honest: no corruption was tolerated. Personal kickbacks and under-the-table commissions were part of most government transactions with the private sector, but SELCO would not play along and thus lost huge potential solar-electrification opportunities. The Indian Renewable Energy Development Authority (IREDA) was another nemesis. IREDA had hundreds of millions of dollars of World Bank funds to develop renewable energy including wind, small hydro, solar hot water, biogas, bagasse (rice husks and sugar stalks to fuel power generation), and, of course, PV. It agreed, in principle, to underwrite agricultural cooperatives that wished to finance SHSs for their members.

I met IREDA officials dozens of times in their modern New Delhi offices and was promised the world, over and over. Harish and I attended their conferences and seminars, and I quickly learned that the entire operation was a smokescreen to protect a bureaucracy that had only two interests: its salaries and the money it managed for the World Bank. The best way to look after the World Bank loan was not to spend or invest it, and the best way to protect your job was not to take risks. Do-nothing bureaucrats are never fired and always promoted. After three years, IREDA finally made a modest $100,000 low-cost loan to SELCO so we could sell SHSs at subsidized interest rates to a spice growers' co-op. The paperwork required was onerous beyond belief: half a foot high, and more than would be required to borrow $500 million from a US commercial bank. I was told by educated observers that SELCO was getting none of this renewable energy money, even though we had almost no competition for it in our sector and were seen worldwide as a visible "success story," because we weren't offering the usual kickbacks, so IREDA just ignored or harassed us. I credit SELCO employees with helping in a large way to redefine business ethics in India.

The Ashden Trust

Dr. Hande, MD of SELCO-India receiving congratulations from Prince Charles for winning 1st Prize at the Ashden Sustainable Energy Awards in London (June 2005).

Meanwhile, IREDA continued to make soft loans and outright grants to solar manufacturing companies that are now out of business, including some whose owners eventually went to jail. The World Bank, of course, was clueless. One way to end your career at the World Bank is to highlight corruption in a member state. (India, to its credit, spends a lot of time root-

Philip Jones Griffiths

SELCO technicians raise a 35-watt pole-mounted solar module at a rural house in India.

ing out endemic corruption; top government leaders from all parties are routinely thrown in the can, and the legal system nails lots of CEOs.)

One donor agency that did come through in 2003 with an attractive financing package to leverage $7 million for subsidized solar loans through Karnataka's rural banks was the United Nations Environment Program (UNEP), with co-funding provided by Ted Turner's United Nations Foundation. Eric Usher, a no-nonsense young UNEP program officer based in Paris, visited me in Washington and Harish in India and then pushed the funding through. Eric later wrote to us: "You have done

Philip Jones Griffiths

more in 12 years than create a great company. You've created an industry."

By mid-2005, Harish and his people had sold and installed over 42,000 solar home-lighting systems in India, illuminating the lives of some 200,000 people (this number will be considerably higher by the

The same view at night: SELCO lights allow children to read at night.

time this book is published). These customers were financed by 435 rural banks, including 150 that had been electrified by SELCO. Not all installations were small 35 Wp household systems; many were 1, 2, even 10 kW systems for banks, universities, schools, temples, mosques, churches, and Christian seminaries. (There are 25 million Christians in South India who trace their faith back to St. Thomas, said to be buried in Madras.) Solar water heaters became a popular product, saving people from having to burn fossil or wood fuel to heat water for bathing and washing. Solar water pumping became a substantial part of the business. SELCO asked Anand Electronics to build solar streetlights, which local *panchyats* (village governments) and community organizations purchased.

Tata BP Solar, under the management of the amiable Arun Vora, an unsung hero of global solar PV production, continued to supply SELCO with solar modules manufactured in its ever-growing factory north of Bangalore (40 percent of its output was for export). Arun, who had identified 25,000 Indian villages that "will never get electricity," was a big fan of Harish and SELCO's mission. Mrs. Rao's Anand Electronics continued to supply its exclusive electronics and lights, while private-label deep-cycle batteries were outsourced and manufactured to the company's specs. SELCO opened a rural training institute at Manipal to train technicians to join its growing staff of employees, which topped 180 by 2004. The SELCO brand became well known, indicating quality, reliability, and service. A young American graduate student interviewed 67 remote customers who all said "they were happy with their solar products." This is a pretty good record for a complex high-tech appliance sold to poor farmers and villagers. (Sorry if this sounds like a marketing brochure, but this is the story of selling solar in the land of the sacred sun.)

India is a world beacon of religious tolerance ... for the most part. Okay, Sikhs killed Indira Gandhi, blew up an Air India 747 over the Atlantic, and one day, when I was in-country, massacred 180 Hindus on a train in Uttar Pradesh, which only rated a small mention in the newspapers. Muslims slaughter Hindus on the slightest provocation, and Hindus respond with massacres of their own, as they did in Gujarat in 2002. In general, though, India's 140 million Muslims and 900 million Hindus get along, intermingling in every town, with the Muslims eating beef in a land where cows are as sacred as the sun. The internal Muslim question is exacerbated by living next door to a near-terrorist state, Pakistan. And of course there is Kashimir.

Then there are the Tibetans. They came over the Himalayas from Tibet in the 1950s, 85,000 refugees and the Dalai Lama, escaping Chinese aggression and an entirely intolerant nation. As an aside, it is interesting to compare, as I have been able to do, two next-door nations with populations of over one billion people each: one deeply spiritual, where religion of one kind or another holds sway; the other mostly atheist and almost entirely without religious values or a spiritual base of any kind. I'll take India any day, for China is heartless and mean by comparison. But China is today the far richer country. We Americans are entwined with both, for China manufactures just about every consumer product we buy, and India answers our telephone calls to banks, credit card companies, and tech support.

Meanwhile, the Tibetans who came to India are not part of any of this, living an isolated existence in rural India. India settled the refugees in ten states, and five Tibetan settlements were set up in Karnataka, which now has the country's largest Tibetan communities. Hindus seem to revere these newcomers, who are spiritual seekers like themselves. There is little possibility they will ever go back to China, and it is equally unlikely they will ever integrate into India. What will become of them is anyone's guess.

SELF's Bob Freling had met the Dalai Lama once in Washington, DC, and garnered support from Buddhists in the United States who were interested in helping the refugees in India. Bob contacted Harish, who visited the Tibetan settlements at Mungod and Bylakuppe in Karnataka. SELF's funds, and subsequent direct orders for solar systems from the Tibetan monks, resulted in the Solar Lights in Tibetan Settlements program that brought

street lighting, temple illumination, solar water heating, and lights for households, monasteries, and the SOS Tibetan Children's Village. This became a
substantial part of SELCO's business in 2001.

The Tibetans somehow collected funds to build an enormous new building in Mungod. The Gaden Jangtse Monastery and Temple covered several
acres. Harish, Thomas, and Pai sent in their best technicians, who lived in

SELCO

The Dalai Lama with SELCO'S
Suresh Salvagi (L) and Nityand
Mukherjee (R) celebrating the
Tibetan temple's new solar
lighting systems.

the community for nearly a year, electrifying the new
monastery, including the Dalai Lama's bedroom and
private quarters.

When the Dalai Lama appeared for the dedication,
he sought out project managers Nitya and Suresh, from

SELCO, and thanked them for harnessing the sun's energy to work for them and for doing such a bang-up job bringing electricity to Mungod. He drew them close, then took their hands in a firm grip and smiled for a photo that now graces every SELCO SSC in India, often mounted right next to the ubiquitous garish calendars featuring all the Hindu gods, who don't seem to mind.

In May 2004 the Congress Party returned to power with a strong rural vote, indicating India's 8 percent growth rate was having little economic impact on 700 million peasants. In some ways, globalization was hurting them as factories were closing, unable to compete with cheaper Chinese goods, even imported silk. Farmers were suffering from the influx of imported foodstuffs. Despite all the talk of high tech, only one million Indians out of a billion (1/100th of 1 percent) worked in IT. And only 659,000 Indians owned computers, less than one one-thousandth of the population. Meanwhile, one of the Congress Party's election promises was "free electricity for rural areas." Electric power was again a paramount issue, but it can never be free. Even "free power from the sun" costs something.

If people want to buy electricity, SELCO-India will be there for years to come, delivering "sun at your service."

# CHAPTER VII

# Sunshine and Serendipity

Serendipity, as you'll recall from Chapter 2, brought me back to Sri Lanka, the teardrop island nation at the tip of India, which I had first visited in 1971. In our late 20s, my first wife and I set off from our base in London with Qantas Airways' round-the-world tickets to see a few selected countries. We chose Sri Lanka, Singapore, Australia, and Tahiti. My wife added New Guinea to the itinerary, so I flew on to Polynesia and stayed at a friend's Tahitian beach house overlooking Moorea. My friend and I spent evenings at Quinn's Bar in Papeete and visited the brand-new airport once a week to look for my wife on the weekly Qantas flight from Sydney. This was before modern communications. Eventually my wife and I became even more out of touch (although she did finally show up in Tahiti); the marriage clearly was not destined to last.

In any case, we did share one wonderful month together on the fantasy island of Ceylon, visiting all the tourist sites with a hired driver: the ancient cities of Polonoruwa, Anuradhapura, Sigiriya; the beaches; the temples; the tea estates. We lived liked sahibs at the Queens Hotel in Kandy, at the government guest houses set in lush gardens with monkeys in the shade trees, and at the historic Galle Face in Colombo, where we were the only guests. The Galle Face was a place to which I was serendipitously to return exactly 20 years later.

Lalith Gunaratne pulled up in front of the now-refurbished Galle Face Hotel on the Galle Road (the coast route to Galle, the Portuguese and Dutch

167

fort city in the south). Built in 1862 in the grand British colonial style, it is one of the oldest hotels in the world in continuous operation. It anchors the Galle Face itself, a huge expanse of open grass south of the old parliament building between the Indian Ocean and the Galle Road. Its rooms are large enough to fly small aircraft in, and ocean breezes make it tolerable when the feeble air-conditioning doesn't. Sri Lanka is seldom as hot as Washington, DC, but its extreme humidity can drive you mad. You can come to terms with it on the long, open verandah beneath a dozen ceiling fans, where white-robed waiters with brass number tags solicitously bring you fresh lime sodas. Several of the waiters I remembered from 1971 were still there in 1991, and one even remembered me. All our solar adventures in Sri Lanka were launched from the Galle Face verandah, including SELCO-Sri Lanka, which I'll get to momentarily.

Lalith jumped out of his red Toyota sedan and came around to greet me on the steps. Built like a wrestler, dark complected, with a full beard, he looked like someone out of a Sinbad movie. He was stylishly dressed and spoke his Sri Lankan English with a lilting accent tempered by his upbringing in Toronto.

"Welcome to paradise!" he exclaimed. "Do you like old car rallies? Samantha and I want to take you with us to the Mercedes Benz rally in Victoria Park."

"Sure," I said, surprised. So here I was on my first afternoon back in Ceylon, now renamed Sri Lanka, and I was going to an upscale Sunday afternoon function with the cream of Colombo society to look at classic Mercedes Benzes. It turns out there are a lot of them in Sri Lanka! I had owned a used Mercedes once, so I could appreciate what I was seeing.

Lalith owned a small, pioneering, solar PV company, which he and two of his fellow Canadian Sri Lankans had started five years before, deciding after college to return to the land their parents had left. Lalith had originally studied "nuclear plant management" at a Canadian university. I asked him later why he switched to solar. He replied, "Because I woke up one night in a sweat with the realization that splitting the atom is an unnatural act." He believed atoms, as the building blocks of the universe, should remain whole. Together with his childhood friends, he had formed Suntec with family investors and a big loan from the Development Finance Corporation of Ceylon (DFCC). Lalith wanted to improve the lives of rural people with solar electricity.

Thanks to the miracle of fax machines, I had been in touch with Lalith regarding Suntec's intention to "sell electricity at the customer's doorstep." I wrote to him that SELF was promoting such efforts wherever we could, and that we could raise money for such projects; we'd already been promised a large sum from W. Alton Jones Foundation to launch a solar project with Sarvodaya, as recounted in Chapter 2. We would need solar home-lighting systems, and I decided we might as well buy them locally, if they worked.

Lalith and his friends had imported equipment from Canada to manufacture their own photovoltaic solar modules. They imported the cells from Japan, which then had to be cut (see Chapter 4), "tabbed," and laminated into 36 Wp modules. Theirs was, indeed, a daring enterprise, since they did not have the volume of sales to warrant local manufacture. It would have been cheaper to import finished modules. They also had problems with unreliable DC light fixtures and their inverter ballasts, which they also made, often with what they didn't know at the time were inferior parts from the local market. Nonetheless, Suntec supplied SELF's two solar projects in Sri Lanka over the years, working jointly with Sarvodaya's technicians — after helping to train them. Through these programs we electrified some 1,000 households, plus 80 Buddhist temples. Everyone, even the temples, had to sign up to pay on installment credit through one of several credit mechanisms SELF launched. Lalith and his two partners, Veren Perera and Pradip Jayewardene (grandson of the former prime minister, Junius Jayewardene), were three of the most extraordinary people I had ever been privileged to meet, all immensely dedicated to this magic technology that could bring happiness to the rural people of this magical isle.

Suntec had a good run (like SELF), and by 2000 the partners had sold it to Shell Renewables, which renamed it Shell Solar Lanka Ltd. Veren

Perera, who looked like a movie star, married a German woman, formed a stockbrokerage, and made a lot of money in the Colombo stock market. He retired at 35 to run a remote yoga retreat for the European back-to-nature tourist trade. Lalith became an international consultant on rural energy, and we continued to cross paths everywhere — at solar energy conferences in

SELCO

SELCO technicians in Sri Lanka prepare to install two 35 W modules atop a tile roof.

Washington, Morocco, Switzerland, China, Vietnam, Amsterdam, and Germany. Pradip stayed with Shell and became the new company's CEO for a time.

Sri Lanka was where I came to understand and to know the rural people of the Two Thirds World. Here I was able to enter this world, their world, the world in which half of humanity lives. It is a world of small farms; of simple, small houses with thatch or tile roofs; and of families usually too large to fit in the usual four rooms. Often these rooms have no furniture and people sleep on floor mats. But it is not necessarily a world of "rural poverty," and the words "rural" and "poverty," I learned, don't necessarily go together. In fact, according to the United Nations, there are more urban poor than rural poor in the Two Thirds World.

The Two Thirds World that I came to know, beginning with Sri Lanka in 1991, is a world of proud, self-sustaining, peasant farmers. There are 700 million of them in India, 800 million in China, maybe 400 million in Africa, at least 200 million in Latin America, and another 300 million in the Middle East, Southeast Asia, Western Asia, Eastern Europe, and Russia. Their lives have not changed much in hundreds if not thousands of years. They mostly own or lease their farms now, as both India and China have instituted vast land reforms, so at least most are not serfs or sharecroppers. They are the majority population of the Two Thirds World, constituting at least half the four billion people in the developing world who make up two-thirds of the planet's total population of six billion. And half the four billion have no electricity; half of all peasant farmers have no electricity. This was the world I came to know intimately, or as intimately as a foreigner can without actually living there.

Until Lalith Gunaratne took me inside a peasant farmer's house, I had never been in one (except in the Dominican Republic). When I first roamed around Sri Lanka in 1971, peasant farmers were mere backdrop, their houses too modest to be worth documenting with color slide film. Now I spent my days in a dozen countries paying endless visits to rural farmers, entering their houses, meeting their families, photographing their smiling, guileless children, patting their wary dogs, and taking tea in what were sometimes no more than wattle-and-daub one-room shacks. But many were more substantial, and even Lalith never ceased to be amazed at how relatively well-off so many rural people actually were, living on single cash crops, a small garden, and maybe some livestock. They often had solidly built houses with hardwood trim, proper windows and doors, red-tile roofs, ceramic floors, handcrafted furniture, running water (hand pumped and gravity fed), and numerous well-furnished bedrooms. But they had no electricity and no hope of getting any until Suntec arrived, and then SELF, Sarvodaya, SoLanka, Shell, RESCO, and finally SELCO-Sri Lanka.

"People in Colombo have no idea people out here live like this. All they think of is the 'rural poor.' Most of our bankers and industrialists have never visited rural Sri Lanka," Lalith lamented during one of our many rural excursions. "They send their kids to school at Berkeley and go on holidays to Europe, but they've never driven across the island to Trincomalee."

This is a world of unspoiled beauty, of peace and quiet, of country rhythms and timeless cycles. There are the quick twilights, the early dawns and clear mornings, and the hazy, lazy afternoons when you only want to rest under a tamarind tree after another noontime platter of red-hot "rice and curry" served on an aluminum plate. The farmers are invariably bare chested, dressed only in a long, colorful, plaid or batik sarong. The women have simple work-a-day saris, while the children are in school uniforms or casual but neat outfits: girls in frilly bright satin dresses; the boys in creased and pleated shorts with white short-sleeve shirts. No one wears shoes, but they have them for going to town.

People in this world are uncomplicated, usually uneducated, although not illiterate, and not naive nor without their passionate loves and hates as they struggle against fate and nature. This was richly portrayed in *The Village In The Jungle*, a classic 1913 novel by Leonard Woolf (Virginia Woolf's husband), which he wrote after seven years in the Ceylon Civil Service. "The villagers belonged to the goiya caste, which is the caste of cultivators. If you had asked them what their occupation was, they would have replied 'the cultivation of rice'," Woolf wrote. "The spirit of the jungle is in the village, and in the people who live in it. They are simple, sullen, silent men. In their faces you can see plainly the fear and hardship of their lives."

The hardship for so many of the rural majority of the Two Thirds World is caused by climate and uncertain weather cycles. No rain means no rice, and they revert to slash-and-burn agriculture, cultivating "chenas" (clearings in the jungle); as Woolf noted, "So sterile is the earth, that a chena, burnt and sown for one year, will yield no crop again for ten years." In Sri Lanka, no rain also means no electricity, for every bit generated on the island is from "renewable" hydropower, which doesn't always renew. But that didn't concern the 40 percent of the population, all of it rural, that had no power anyway. No, they were concerned with eating and with feeding a growing population on limited lands in an equatorial climate where soils are notoriously poor, where jungles grow and crops do not.

But despite the fragile existence of this ancient rural life, what I saw was "sustainability" in action, a work ethic unknown in the urban world, and a self-sufficiency that is unimaginable to most of us. (Lalith and I believed we could help make them SELF-sufficient in electricity, too.) How did the best-off among them pay for their nice houses, furnishings, motorbike? Usually someone worked in town or in Colombo or abroad. Many young mothers worked as maids in the Gulf States; many a husband in Saudi Arabia or Oman. Rural life wasn't as bucolic and content as it sometimes seemed, because the people had very little cash; sometimes they used half their cash income to buy kerosene for lighting and cooking, salt, and cooking oil, while rice and vegetables were cheap and plentiful.

Rural life nearly came undone in Sri Lanka in the 1980s when a Singhalese Marxist peasant movement, the JVP, emerged to challenge the state. Not unlike the Khmer Rouge, or the Shining Path in Peru, the JVP was led by an ardent "rural revolutionary," who thought the peasants in Sri Lanka were getting short shrift. This is the other, not so pretty, side of the rural Two Thirds World. Because people have little access to information, with few newspapers — and no TV because of no electricity — they begin to feel isolated and left behind by a world they can't be part of and don't understand. Young people in these rural areas are easily mobilized by a charismatic leader; an old story, perfected by Mao Tse Tung, whose uprising succeeded.

The JVP's uprising didn't succeed in Sri Lanka. By 1989 its revolutionary leader had been captured and killed by security forces. Global terrorism expert Rohan Gunaratna told me over lunch on the Galle Face verandah how this organization nearly shut down the country; worse was the government's response. The story remains nearly unknown in the West, but to root out the JVP, government forces are said to have arrested and killed thousands of teenage boys — no one knows exactly how many — from villages throughout Sri Lanka. Whole classes of boys were found beheaded on their school grounds. The government felt it had no choice. Gunaratna's book, *A Lost Revolution*, describes the terrors to which every Sri Lankan, rural and urban, was subjected during those dark years. I had arrived in Sri Lanka a year after this madness ended, and Rohan set me straight about a period of Sri Lankan history no one wanted to talk about. (*Anil's Ghost*, a novel by Sri Lankan-born Michael Ondaatje, author of *The English Patient*, deals with the aftermath of those years.)

This shows why rural development is so important. It is critical to bring some basic services to the people living on the land. They will be better served by staying on the land, as opposed to joining a peasant revolution or moving to a city, where they will end up in squatter shacks and in slums. SELF's funders shared this view, and later we found investors who saw delivering solar power services to rural areas a signal opportunity for "responsible capitalism," and not just a social mission.

Others saw it as sound international policy: electricity as an entitlement. One member of the European Commission told me in Munich, in terse terms, "If we don't bring electricity to them there, they'll all come here." And coming to Europe they are. Sri Lankans, Bangladeshis, and Indonesians, as well as North Africans attempting to "immigrate" to Europe, regularly wash up on the beaches of the Mediterranean. The European authorities only find the dead ones.

The rural people I came to know as I was invited into their humble homes were decent, hopeful, generally happy, friendly, outgoing, and extremely capable at cultivating whatever crop they chose. Development experts claimed they needed clean water, but they all had wells or access to good village cisterns. The donors' consultants said they needed medical care, not electric lights or TVs, but they all had access to numerous rural clinics. The international agency officials said they needed housing first, but nearly everyone I saw already had a nice house, with a neatly swept courtyard, flower gardens blooming profusely, shading palms, and food often dropping from the trees.

What they *didn't* have was electricity. They couldn't afford batteries to run radios all the time; they tried to watch small 12-volt black-and-white TVs by charging car batteries, and recharging them was a common local enterprise — the batteries were carried on bicycles to the nearest town, a big pain. And their children, whom they cared about most of all, could not study well by kerosene lamplight. Their kids did not wish to emulate Abe Lincoln, not when they knew something better existed, which Abe didn't have: *electric* light.

For the next ten years I commuted to Sri Lanka from Washington, DC, as we sought to bring solar rural electrification to this ill-fated if beautiful island. Unserendipitiously, the year after the government put down the JVP insurgency, the Tamil revolt erupted. The Liberation Tigers of Tamil Eelam (LTTE) sought self-rule for the north end of the island and the city of Jaffna. LTTE's leader was the self-declared revolutionary Prabakaran, a monster who had ordered the assassination of Indian prime minister Rajiv Gandhi by a young, female suicide bomber, as mentioned earlier. (By 2005, Prabakaran had ordered 220 suicide attacks.) I scribbled in my diary in 2000, "Prabakaran is the most ruthless and violent terrorist running loose on the planet. He makes Bin Laden and his terrorists look like rank amateurs." Bin Laden would catch up a year later, and Americans would know what it is like to live in a state of terror.

Over the decade I worked in Sri Lanka, the violence got worse and worse, but it seemed to have no effect on the good life in Colombo, on our rural projects or our company's operations, nor on our ability to travel around nearly all of the country. The war went on like an argument in the next room to which no one paid attention, except the rural families of the 64,000 government soldiers, Tamil Tigers, and civilians who had died. The news media barely paid attention to a war that consumed thousands of young soldiers every year; reporters were not allowed near the battles. Governments came and went; one prime minister, Premadasa, along with 24 of his aides, was blown up by female suicide bombers at an election rally (that democracy works at all in Sri Lanka is a miracle). Tiger terrorists blew up the State Bank in 1999, killing hundreds, then set off a huge truck bomb in The Fort, the financial center of sprawling Colombo, and blew out all the windows of the Intercontinental Hotel, the Hilton, the Meridien, and the two 39-story towers of the brand-new International Trade Center.

Prabakaran's Tigers had nothing to do with redressing rural poverty; it was a vicious thug's personal fight, nothing else, hiding under the guise of demanding ethnic rights. Tamils, who'd been in Sri Lanka for 1,000 years, brought in from South India by the Singhala kings whose people didn't like to work, comprised half the population of Colombo, owned most of the large businesses and industries, and had a progressive determination that sometimes seemed lacking in the more laidback Singhalese. A famous Tamil family owned the Galle Face Hotel (maybe that's why I always felt safe there). But Tamils in Colombo had little or nothing to do with the Tigers. Neither did the Tamil tea pickers in the hills, brought in by the British; they didn't support what foreign supporters claimed was an ethnic uprising, and they weren't violent. Most good citizens of Sri Lanka recognized the LTTE for what it was: a demented, twisted, and deadly terrorist organization built on the backs, and lives, of brainwashed youth.

The Norwegians were everywhere, trying to negotiate a peace accord as they had done with the PLO and Israel, but they didn't have much luck. I sat next to the Norwegian ambassador on one flight out to Europe, and he told me he'd just come from a meeting with Prabakaran (I knew, and asked him about it, for it was in the papers). He said they were ready for peace. I thought him naive. Two weeks later, in 2001, Tigers blew up nine jet aircraft on the runway at Bandaranaike International Airport, destroying military and civilian planes, including two of Air Lanka's brand new Airbus 340s. Twenty Sri Lankans also died in the blasts, but the Tiger sapper squads took care not to kill a single tourist (unlike the mad Islamists in Egypt, who had just gunned down 59 tourists at Queen Hatshepsut's Temple at Luxor). Tourists were *never* at risk in Sri Lanka, despite State Department warnings that kept virtually all Americans away (in ten years I never met a single American tourist in one of the most glorious tourist destinations on the planet). The European package tours continued to arrive from Leeds, Sheffield, Dusseldorf, and Amsterdam, ready for sun and beach and clueless that there was even a war going on. For them, Sri Lanka was nothing more than a cheap beach. I always felt safer in Sri Lanka than in Washington, DC.

It was the economy that took a beating from the endless war. And this helped no one, least of all the "rural poor" who were the focus of our business ventures.

Priyantha Wijesooriya, mentioned in Chapter 6 as the graduate student who introduced me to Harish Hande, also led me into business in Sri Lanka. He came to the verandah of the Galle Face one day in 1996 with his wife and daughter and brought me a business plan for the SoLanka Electric Company. The nonprofit SELF had launched the commercial company SELCO in India, and now Priyantha, whom I'll call PW, wanted to convert SoLanka, an NGO we had formed earlier with SELF's support, into a business operation. He later renamed his venture RESCO, which stood for Renewable Energy Service Company.

PW was finishing his PhD from the University of Massachusetts, following on from his work at the London School of Economics, and he was certainly qualified to take this next ambitious step. I had enjoyed working with him on pilot projects, such as the solar electrification of Morapatawa village, where he had raised sustainability to a high art. This was a very poor community of cashew growers north of Puttalam on the northwest coast, truly a village in the jungle, accessible only by five kilometers of dirt track. Here we had installed 150 Suntec SHSs, set up a local revolving fund with village management, and PW had even opened a solar-powered "jungle workshop," where technicians he'd trained began assembling and soldering together their own charge controllers and luminaires (DC electronic light fixtures). They also learned to install and maintain the systems. This became one of our truest models of SELF-sufficiency, visited by endless delegations of consultants, rural aid officials, and graduate students. Mr. Rupasinghe headed up the solar committee of Morapatawa village, and Mr. Goonesekera was its secretary/treasurer. Year after year they diligently collected the monthly installment payments from each family, oversaw the maintenance of the systems, and reported regularly to PW in Colombo. With the funds collected, they began to make more loans to more families to buy SHSs.

PW was a strong force of muscle, will, and visionary brilliance, like Lalith, and I introduced the two of them one day early on, for they would have to work together. They became friends, but remained in their separate spheres: Lalith's Suntec manufactured and supplied SHSs, while PW sold and installed them through our SELF pilot projects — but not in the quantities we'd hoped for.

Morapatawa, and several other nearby communities with solar projects supported by SELF and Rotary International, were not enough to change the rural landscape from darkness to light. PW recognized this, and so he proposed setting up RESCO as a semi-public company, with my help.

I wasn't quite ready to make the jump from rural development to commercial business, even though we had already done it in India. Instead I encouraged PW to carry on without me (which he did, as you will see), and I concentrated my efforts on securing some large infusions of philanthropic funds from the United States for our Sarvodaya project. Dr. Ariyaratne, the most famous person in Sri Lanka (he declined many offers to run for prime minister), had decided to put the full force of his huge NGO behind the

SELF/Sarvodaya Solar Project. Sarvodaya and its rural credit banks had access to 8,000 villages, and I believed this program could really make a difference and bring solar lighting to rural Sri Lanka.

Lal Fernando, the chief of Sarvodaya Rural Technical Services, was named the solar program manager, and I introduced him to Lalith and the Suntec boys. I also brought Lalith in to meet with Dr. Ari at his offices in Moratuwa. Lal was a dedicated and professional engineer with a decade's experience handling Sarvodaya's rural water projects. He decided to take on solar electrification as well. A devout Catholic, he nonetheless strongly believed in Dr. Ari's concept of "right livelihood for a full engagement society" and Sarvodaya's Buddhist ideal of "universal awakening," and he worked tirelessly for this great "people's movement", crisscrossing the highways and byways of rural Sri Lanka in his Toyota double-cab pickup.

Lal introduced me to Saliya Ranasinghe, a former banker who now headed up the Sarvodaya Economic Enterprises Development Services (SEEDS), the huge island-wide credit facility we would rely on for many years to provide credit to rural householders so they could purchase solar-electric lighting systems. SEEDS' banking division had enrolled nearly half a million members of local Sarvodaya Shramadana Societies (self-reliance groups), and was providing microcredit to the poor long before it became fashionable in donor circles. Saliya was a large, blustery, rambunctious man who had left commercial banking to dedicate his life to the Sarvodaya mission, and he was a serious professional with an appetite for challenges. I had to work hard to sell him on the solar loan program that Dr. Ari and I proposed he set up. But when he got into it and understood the ramifications of energy self-sufficiency and how Sarvodaya could be the world's first leading player in delivering it, he got very excited.

By the late 1990s, SELF, Suntec, Sarvodaya, and SEEDS — and PW's independent RESCO-Sri Lanka — were cooperating to bring solar lighting to some 2,000 families scattered from Kurunegala in the north to Galle in the South. Sri Lanka was becoming the world's crucible for solar rural electrification.

After Anil Cabraal and I led a World Bank mission to Sri Lanka in 1993, as described in Chapter 3, Anil, the Sri Lankan-born solar energy expert at the Bank, and I convinced the Bank's Asian Alternative Energy Unit to propose the $32-million Sri Lankan Energy Services Delivery Project (ESD) to the government of Sri Lanka and the World Bank board of directors. The ESD project was not approved by the Bank's board until late 1998. It took six years to develop a project that any private corporation could have put together in six months easily.

PW wasn't waiting. He had already found the progressive Dutch bank Triodos and its socially conscious investment arm, Hivos, were interested in his RESCO business plan. PW closed down his NGO, SoLanka, and launched RESCO-Sri Lanka as a semi-public company, approved by the country's board of investment, with funding from the Dutch, the People's Bank of Ceylon (the largest government-owned bank), and SELCO. We spent one day in the paneled offices of the People's Bank chairman, overlooking Colombo, as we sat around a huge table signing endless documents, shareholder agreements, bylaws, certifications, and various government papers officially creating the company.

The deal had been consummated by Nissanka Weerasekera, a young investment banker working then for People's Venture Investment Company, who had previously worked for USAID in Colombo, which is where I had met him. Nissanka was the kind of person to give one hope that hard times were not going to hold Sri Lanka back. Thoroughly "Westernized" and well-educated, with a brilliant wit, he could have emigrated to the States at any time, but he chose to stick it out at home. Even after a suicide bomber nearly killed President Kumaratunga at a public event near Nissanka's daughter's elementary school, killing 18 people, he drove his daughter back to school the next day and gave no thought to leaving paradise. We named Nissanka to the board, along with Hans Schut, an officer of Triodos bank, PW, and me. I was named chairman, and PW was chief executive.

We held our first board meeting swimming together in a cool, clear, jungle stream near Wellawaya, a provincial town where RESCO set up its first solar service center.

I was now chairman of two commercial companies in two countries, inventing this new reality, which consisted of cash flows, balance sheets, income statements, loans, debt, interest rates, personnel, and of course supplies, materials, import duties, marketing, sales, receivables, and all manner of technical issues, as best I could. We imported PV modules by the half-container from Siemens Solar, Solarex, and BP Solar. Fortunately, by this time computers and the World Wide Web had replaced fax machines and I could play my role from Washington, effectively interacting in nearly real time with Colombo and Bangalore (and Ho Chi Minh City, as you'll see in Chapter 8). I was always 11, 12, or 13 hours out of sync, however, as their day is our night.

In order for this business to work, people had to be able to "access credit." But first they needed the credit to access. Rural banks in Sri Lanka continued — and continue to this day — to refuse to take risks on rural people without sufficient collateral. In contrast to banks in India, they would not take the SHSs our customers purchased as security, even if we offered to send out solar repossessers to settle unpaid loans. So we turned to Saliya Ranasinghe, SEEDS, and the Sarvodya Shramadana Movement, which by now had financed nearly all the solar lighting systems sold on the island to date. The legions of World Bank credit "experts" independently came to their own conclusion that only Sarvodaya had the "capacity" to become one of the Bank's "participating credit institutions." All my early work with SELF, Sarvodaya, and SoLanka had paid off, and the first World Bank solar rural electrification project got underway. A person had to be a member in good standing of the Shramadana Society to secure a World Bank-subsidized ESD loan through SEEDS, and by 2000, Sarvodaya had financed 6,000 SHSs.

But then, just when success seemed to be at hand, the problems started. Everything moves very slowly in the developing world — maybe that's why it's still developing. Time is different there, palpably so. As the Sri Lankan supposedly said to the Mexican, upon learning the meaning of "mañana" (which famously expresses that you can always do tomorrow what you didn't do today), "Sir, we have no word in Sinhala to convey such alacrity."

Tomorrow means next month on this languorous, steaming isle; a month means a year. What we in the West are accustomed to doing in hours takes many, many days in these unhurried lands. Time is culture, and they have plenty of both.

This meant that getting paid by the World Bank's ESD Project, the funds of which were managed by the State Bank and the aforementioned DFCC, would be done on Sri Lankan time, thank you. Not World Bank time, not American time, not SELCO time, not according to business standard time. Credit only works if the lender pays the supplier for the goods purchased by the borrower. Of course you must allow time for a loan to be processed, especially one that originates in a rural area on paper, but 30 days should be sufficient. In the West, 10 or 15 is normal for credit transactions on paper. Merchandisers and retailers get paid by credit card companies almost instantly now, as it is all electronic. The entire Western economic system, which has built the mightiest economies ever known, is based entirely on the ease and efficiency of credit. America borrowed its way to greatness. Not so the Two Thirds World.

Since most World Bank economists have never run a business, economic theory must suffice. And so the legions of Bank economists who streamed in and out of Sri Lanka during those years to prepare our solar finance program managed, in the end, to come up with a problematic project that was deformed at birth, which could work only in theory. Yes, it was funded, and yes, $32 million was now available for long-dreamt-of credit for farmers' "solar loans." But then intractable local bureaucrats at the government "apex lending institutions" were put in charge, not private commercial bankers. These intransigent fools made Indian bureaucrats look efficient. And, strangely, unbeknownst to the entrepreneurs taking huge risks at their new solar enterprises, the solar project had been designed to eliminate all risk to the World Bank. Risk was forced downward, landing exclusively in the laps of RESCO, Suntec, and two other small solar companies the program had spawned.

Nonetheless, the national solar program took off, and soon enough we were signing up farmers as fast as our fleet of motorbikes could carry our salesmen down every country road and up every jungle track. We would bring along the local Sarvodaya credit officer from one of the 2,600 Shramadana Societies, which enrolled members in the SEEDS program. The salesman would pitch our solar lighting system, and the farmer would apply for credit over tea in his house, and an SHS sale would be "booked" in a subsequent visit, followed by installation a week or two after. We had our sale, the customer was thrilled to have solarelectric power and light, and Sarvodaya had the loan papers, which now had to make their way through an impenetrable maze of steps that had been concocted by the visiting experts to avoid risk. And by the end of the steps, what was a loan to the government of Sri Lanka at International Development Agency (IDA) rates of 1.5 percent became a 16 percent loan to our customers after each lending agency along the way — State Bank, DFCC, and SEEDS — took its quotient of "points" for risk.

Soon I learned to my chagrin that our Sri Lankan solar company wasn't getting paid in a remotely timely manner by the World Bank's illustrious ESD solar program. Receivables were running to 180 days. And this was when the program was running with alacrity. It's impossible to operate a sustainable and growing business this way, but that didn't matter to the World Bank crowd, which flew around the world giving PowerPoint presentations at development conferences on their successful solar project in Sri Lanka.

Because I was in Washington, I could not oversee the day-to-day operations of our companies, nor did I want to. But our managers in the field were often intimidated by the World Bank consultants who continually parachuted in to supervise and "monitor" the project, and by the arrogant local project staff of the Bank, who held the future of these companies in their hands. It was not a happy marriage of public and private interests for the common good that we had idealized might be possible. But we nevertheless struggled on year after year.

It became clear that my friend PW — now Dr. Wijesooriya — was not the world's best manager, and he was soon hard-pressed by these cash-flow and credit concerns. John D. Rockefeller once said, "You can build a friendship on business, but you cannot build a business on friendship." How right he was. PW was my friend, but he was putting our entire investment at risk.

So in 2000, SELCO bought RESCO by purchasing the minority shares of Triodos and People's Venture Investment Corporation. We renamed the company SELCO-Sri Lanka, changed all the signs and logos, and found a new manager, relieving the beleaguered PW to retire to academia. The company would not have existed without his tireless promotion of his dream, but it was time for him to go. I was the last to see it, but when I finally did, I took immediate action. He took it graciously, and we remained friends.

Harish Hande, from India, had been helping out in Sri Lanka. I had also "lent" RESCO Kamal Kapadia, the Oxford graduate intensely interested in rural development, who was introduced in Chapter 6. She was on the American SELCO payroll, but wished to remain in the field, and she traveled by herself around rural Sri Lanka, training local youth in business administration and generally helping out with marketing. There was nothing she couldn't do, including, along with Harish, finding a new manager for our operations in Sri Lanka.

They discovered Susantha Pinto managing one of Sri Lanka's largest tea plantations and tea factories with 1,400 employees. We had electrified some of the "line houses" belonging to the Tamil tea pickers, with the tea company guaranteeing the loans, and Susantha had quickly noticed that solar energy was an interesting idea and more fun than growing tea!

Before he was 30 he had reached the top level in the tea business, superintendent. He had no formal education, but had done agricultural studies in Germany and Thailand. He wanted a change. I interviewed him in Sri Lanka at the Galle Face, our Sri Lankan board of directors met with him, and we offered him a job as CEO of SELCO-Sri Lanka.

Susantha was a bear of a man, used to picking up recalcitrant tea workers with one hand and dropping them, not in anger but to make a point, such as "I'm the boss." He oozed power and charisma and he was unflappable under pressure; nothing could rattle him. He was also capable of getting six

jobs done at once. I had never met anyone in business anywhere with as much energy and drive. No one needed to tell him what to do, and so I didn't. I asked Harish only to give him advice. We announced the creation of SELCO-Sri Lanka at a well-attended press conference at the Trans Asia Hotel on May 23, 2001. Susantha quickly ramped up sales to several

The author and Susantha Pinto, SELCO-Sri Lanka's CEO, unveil the SELCO logo, establishing the brand on the island nation.

hundred SHSs a month. Kamal stayed around to help him launch the new venture before she headed off to Berkeley to work on a PhD.

Susantha was always very formal, insisting on calling me Mr. Williams. I never could break him of the habit. He also dressed formally, with a crisp suit and tie, despite the withering humidity of Sri Lanka. His employees also wore ties, even at the service centers, except for the technicians, whom he put in bright SELCO uniforms with embroidered logos on their shirts and hats (women, of course, as in India, wore saris).

I was always amazed at Sri Lankans' formal business attire, even in our small solar business, and I had to do as the Romans and wear a tie myself everywhere. I loved to joke with the casually dressed managers of SELCO-India that if they didn't shape up, SELCO would order a new dress code based on Sri Lanka. South Indians *never* wear ties. Funny, the cultural differences of two subcontinent nations and ethnic cousins separated by only a few miles of water. (A further note on apparel: Indians, Sri Lankans, and Vietnamese, men and women, always took pride in their appearance, always looked neat, freshly pressed and clean, putting Americans and even Europeans to shame for their slovenly dress, which has somehow become standard today at home and around the world. Rule one: No one wears shorts in India or Sri Lanka, but everyone is too polite to notice when the foreigners do. I wore them initially to the field, until the mosquitoes got me and I contracted mosquito-borne filaria, or elephantiasis, which fortunately is curable. I stuck to long pants after that.)

I returned some months later, after we'd capitalized SELCO-Sri Lanka sufficiently, and we inaugurated a new three-story headquarters building in suburban Colombo. Arthur C. Clarke, who had been knighted by Prince Charles a few months earlier, accepted an invitation to speak at our event, but security for a simultaneous visit to the capital by China's premier Jiang Zemin closed down all the roads in the capital, and he couldn't get there. Afterward, Susantha, Lalith, Kamal, and I trooped over to his house for an audience with the great man, whose Arthur C. Clarke Center at the University of Moratuwa had been researching PV for years. Sir Arthur had been a long-time supporter of Lalith's Suntec company.

In the early 1990s, Arthur had agreed to narrate the opening of a video, "The Eternal Flame," on SELF's projects in Sri Lanka. Willie Blake (not your usual tongue-twister of a Sri Lankan name), Lalith's uncle, who earned his film stripes as a second-unit cameraman for David Lean's "Bridge Over the River Kwai" (filmed in Ceylon), shot the video for us. In his introduction, Arthur says, "For the last few centuries since the industrial revolution, we've been living on our capital, on the energy stored up in coal and oil from sunlight hundreds of millions of years ago. We've been eating up these reserves at a colossal rate. In the near future, maybe only a few decades, the oil and gas and coal will all be gone. That means we'll have to go back to the original source, the sun."

In 1999 Arthur told the *New York Times*, "There are so many things coming to a head simultaneously. The population. The environment. The energy crunch. I feel rather depressed ... I think we have a 51 percent chance of survival. I would say the next decade is perhaps one of the most crucial in human history."

We're in that decade now, climate change and global warming are upon us, oil prices are skyrocketing as production begins to reach its final peak, and the Middle East is in flames. And this is not science fiction. Yet solar energy was still science fiction in the United States, where Republican Senate majority leader Trent Lott was calling it "hippie technology." At least in the rural areas of Sri Lanka solar power was a reality as we entered the new millennium. It represented the future, and it gave us hope.

More importantly, solar power gave rural people hope. As I've already mentioned, the greatest joy in this business has been visiting the homes of local farmers to witness their own joy at finally having electricity in their home. Over the decade of the 1990s I traveled thousands of miles on Sri Lanka's back roads with Lalith of Suntec, Lal Fernando of Sarvodaya, Priyantha of SoLanka, and Susantha of SELCO-Sri Lanka, and at every house we visited I'd ask the residents how they liked their SHS. My colleagues would interpret for me. At one well-crafted stucco-and-tile house deep in the jungle, Mr. Sugathapala told me, "We were fed up with lighting kerosene lamps and carrying the battery to 'charge.' Today we have good lights to study by. Since there is no soot, the walls are nice as well. We are very pleased that SELCO was able to provide new light for our home."

A neighboring woman, Pushpa Samarawickrama, told me, "Solar power has changed everything for us. My children can now study late into the night, while there are no worries about bottle lamp accidents. Most of the villagers are involved in self-employment projects such as farming, livestock, and sewing. Earlier, we could not work after 6 PM as it was dark. But now we can continue our work till late. What is more, this also drives away the wild elephants who used to roam at night. We have sunlight whether it is day or night!"

One farmer showed me his four bare rooms, two with sleeping mats, one for the kitchen, and the fourth for living and eating. There was no furniture. But there were light switches embedded in the walls, and he proudly turned on the lights, room by room, and said, "The last thing a farmer needs after coming home from the fields at the end of a long day is to fumble around in the dark trying to light a kerosene lamp. Now we just switch on the lights, like in the city!" His wife added, "Our solar lighting unit has been a great boon to us — it is especially useful late at night in emergencies when all we have to do is flick a switch."

This was our company's bottom line, right here; these were our real profits, and the profit to the community. It was most rewarding when I encountered whole communities that had elected to go solar. This meant telling the local politicians, who had promised electricity but not delivered it, to go to hell and not bother them anymore about grid power. Usually these efforts, in Sri Lanka, started with the village temple — specifically with the local priest, who in Theravada Buddhism is as much a social worker as a spiritual leader. Buddhism considered the sun a deity, and Buddha had said, "We give salutations to him, who is the sun, from where all golden rays emanate to light up this living world." We needed to get the Lord Buddha involved in our marketing program; he understood!

Suntec learned to electrify the temple and the priests' quarters first — often as a gift — and the village would follow. SELF had electrified, as already noted, about 80 temples, but the temple societies had to pay, and they did, willingly. One priest, Venerable G. Dharmaratana Thero, said, "Before SELCO's arrival in the village, everyone used kerosene lamps and was facing many hardships. The SELCO systems have been immensely beneficial to carry out the religious and social activities of our temple." Evening activities in the temple courtyard improved markedly when the solar lights drove out the venomous snakes that formerly lurked there. I always got a kick out of how delighted the saffron-robed priests were when they could watch TV, which the temple committee usually bought for them once they got solar power. And after the community temple got lights, every villager wanted them, and SEEDS' loan officers could not sign up people fast enough.

One day I got drafted to help dedicate the foundation of a new temple near Kurunegela. "Come. Come with us," I was entreated after we had inaugurated a solar-powered school for Sarvodaya. I thought they wanted me to watch their ceremony, but instead they wanted me to lead it.

"It is propitious for the temple dedication to be done by a distinguished visitor. You are the first American to ever come here," I was told by a young man who seemed keen on my participation.

"What do you want me to do?" I inquired as we marched through the coconut palms to a clearing.

He introduced me to the young priests, who smiled at their catch — a foreigner to help dedicate their temple foundation! "Here, hold this on top of your head," they said and handed me a small bowl containing lotus leaves and burning incense. I did as told. About 100 villagers fell into line behind me while the two priests led the way. Four men carried poles that held up a large, square, silk banner forming a fabric roof over me. Everyone had flags; incense was burning on top of my head. Slowly we moved along the path to the temple site a quarter mile away. I felt like an idiot. Were they making fun of me, or was this serious?

At the site, where survey lines had laid out the new temple's foundation, I was asked to lay the first brick. This I could do, since my grandfather was a bricklayer and I had learned at his hand. But it was only one brick. By this time I was ready to lay up a whole wall. I made a note to remember to come back and sell the temple society a solar lighting system when the building was finished.

The young technicians we trained were quick to learn, and they should get the largest share of credit for making solar electrification possible. Lalith and Priyantha personally trained dozens of technicians, many of whom became trainers themselves. Training courses, refresher retreats, and technical seminars were always underway somewhere to meet the need for skilled workers who could install SHSs in rural homes.

One of these was Chandral Chandrasena, a Sarvodaya Rural Technical Services employee, who quickly took to the technology. Chandral had an aura of distinguished contentment and a deliberate demeanor that focused him intently on his work of "right livelihood." He drove his Sarvodaya van over the roughest roads, to the remotest houses, and in the midday heat he climbed through the rafters, stringing wires, and onto thatched and tiled roofs to mount solar modules. The work was both mental and physical, requiring the skills of an electrician and the balance of a trapeze artist. The hours were long, six days a week, and month after month Chandral and his crew installed two SHSs a day for Sarvodaya. After each installation was complete, he patiently trained the appreciative family in its use. Chandral never complained or made a fuss. As he was employed by Sarvodaya, his pay was minimal, but that didn't matter. He told me he was doing "God's work" and it brought him deep satisfaction. "Installing solar systems is a labor of love," he told me one day. "I like to share the joy and excitement when I see the village children experience electric light for the first time."

I always felt humbled by such examples, and amazed by people with this much dedication and principle. All the high-tech solar modules, all the sophisticated electronics, all the detailed financial planning and well-crafted business models meant nothing without a Chandral to install the apparatus in a rural house and show the family how to operate it. Where did these people come from? Sri Lankan, Indian, and Vietnamese culture fortunately produced many of them, many Chandrals, at least in places not yet contaminated by greed, consumerism, apathy, despair, and the suffocating demands of finance capitalism. Our solar systems brought hope to installers and customers alike. As Dr. Ari had written regarding Sarvodaya's self-help community projects, "We make the road and the road makes us." The same could be said for solar. Here was something worth doing, and worth doing well, and I always marveled at how well these guys — and some girls — could do it. Chandral himself later took a job with Shell Solar when we failed to hire him at SELCO, a loss for us.

What I learned in business in Sri Lanka, as we sought to explore "new economic paradigms" and socially responsible business models, is that you get the best results when people are motivated not by money, but by something intangible, such as a love of service or a passion for the cause. This, which I call the Third Way, was best described by the British business writer John Kay: "Individuals who are most successful at making money are not those who are most interested in making money." He cites numerous corporate case studies to prove it. I thought of the late Mohan Singh Oberoi, founder of the luxury Oberoi Hotel Group in India, who said: "You think of money and you cannot do the right thing. But money will always come if you do the right thing. So the effort should be to do the right thing."

This was SELCO's credo in Sri Lanka, in India, and in Vietnam. It was too bad that some of our shareholders (or our shareholders' financial managers) and even a few members of our own board of directors in the United States could not understand this. I was to learn the hard way that "the rights of capital" and "doing the right thing" are not always reconcilable or compatible, at least not in the eyes of arrogant young investment analysts and portfolio managers who learned everything there is to know in the world from business school. We knew we had to be economically profitable — it went without saying, and I loudly preached profitability over social service whenever I could — so it made me crazy to think about the donor agency bureaucrats who spent their every waking hour trying to figure out how to put us out of business. (There were exceptions, of course, and you know who you are.)

At the dawn of the new century SELCO-Sri Lanka had 180 employees and was selling and installing 300 or more SHSs per month, sometimes

SELCO

A typical SELCO solar service center in Sri Lanka, with delivery van and salesmen's motorbikes. The center has grid power with solar backup.

hitting 500. Given that each sale was equivalent to an American family buying a car, this was a substantial business for rural Sri Lanka. The company had 10 solar service centers around the country and over 100 dealer-agents selling SELCO SHSs. Susantha was a celebrity, appearing on TV and in the newspapers and business magazines. However, the company continued to be plagued by its long-term receivables and operating debt.

Being the pioneer doesn't always guarantee success, for now we also had imitators and competitors. Shell Renewables had bought Suntec, renaming it Shell Solar Lanka Ltd., and under the able leadership of Damian Miller, who had learned everything he knew about solar from me at SELF and from studying SELCO-India on his MacArthur Fellowship, Shell soon had as many workers, service centers, and motorbikes in Sri Lanka as SELCO. Shell took half our market share. Lal Fernando told me, "Shell selling solar is like an arms dealer running a refugee camp." Well, maybe. I never knew what Shell's game was, except that it spent tens of millions of dollars advertising its interest in solar energy, with centerfold layouts, featuring PV modules glinting in the sun, in *Time*, *Newsweek*, *National Geographic*, and just about every other glossy national and international magazine. The consolation for us in Sri Lanka was knowing that the World Bank program would not work — and was not set up — for just one company. And a business without competition is not a real business. So now we were real, thanks to Shell.

Also very real were the cash-flow problems resulting from the lassitude of the bureaucrats, in Sri Lanka and Washington, in charge of the World Bank ESD program. Shell had deeper pockets and its own supply of solar modules, so it could weather the receivables problem more easily than we could. Our auditors, Ernst & Young, one day reported that they were going to "write down" about $150,000 worth of receivables owed us by the World Bank as bad debt. "The World Bank as bad debtor!" I howled back in Washington, where I promptly called down to the Bank on Pennsylvania Avenue. I got hold of the people in charge of the Sri Lanka energy program, who by chance were hosting the most obtuse Sri Lankan bureaucrat at their offices. He was reporting to them just how well the solar program was running.

I told a different story, and three hours later all three World Bank staffers, including the Sri Lankan manager from DFCC, were sitting around SELCO's conference table at our Chevy Chase offices, trying to explain why the World Bank couldn't pay its bills.

"How can we run a company like this?" I asked them. "And how can you continue to allow this to go on?"

"We're just following bank regulations and procedures," said one of the young economists.

"I can't hold Sarvodaya responsible any longer. They are processing our customers' loans now in 30 days, but it takes them months to get paid by DFCC, so in the meantime they can't pay us," I said, frowning at the Sri Lankan official.

The DFCC manager looked over his glasses at the young economists and complained, "The State Bank sometimes won't release the World Bank ESD funds to us and then we cannot pay Sarvodaya."

He was passing the buck, and the bucks weren't getting passed to us. I was promised this would get fixed once and for all. But nothing was done, and at the end of the year E&Y made us take the write-down for unpaid receivables from the World Bank project. Several years later the cash-flow problems remained unsolved, thanks to many layers of buck passers, and they threatened to spiral out of control, requiring infusions of new capital.

Nonetheless, despite all the financial challenges, SELCO-Sri Lanka struggled along somehow, even profitably during some quarters, continuing to deliver rural electricity on a lush, mountainous, tropical island where 70 percent of the population is going to remain in the dark unless they hook up to the sun.

# CHAPTER VIII

# War Is Over If You Want It

War is never really over — in spite of what John Lennon and Yoko Ono proclaimed from hundreds of billboards around the world in 1969 — but the Vietnam War finally ended for me in 1993, when I first returned to the land of our 14-year "conflict."

Memories flooded back when I visited the Museum of American Aggression in Ho Chi Minh City, as Saigon was now called, and revisited Chivral, the downtown coffeehouse where GI's and journalists used to buy marijuana cigarettes, perfectly rolled and wrapped in cellophane-covered Salem packs — cartons of them. Across from Chivral, the old Continental Hotel had become a four-star property with a fine Italian restaurant on the verandah, and opposite, the old Caravelle was being turned into a spectacular luxury hotel and condominium tower. The rooftop Caravelle Bar still offered the same panoramic views of the former Paris of the Orient: the broad boulevards, the Saigon River, the brick cathedral, the opera house below. From here, as a young freelance correspondent, I had watched the phosphorous flares illuminating the city's perimeter at night and felt the low rumble of B-52s unloading vast quantities of bombs on the delta. Now there were tall office towers sprouting up, including the Citibank building across Le Loi Street, where we opened our SELF and later SELCO accounts.

In 1968, with 535,000 Americans in-country, the war we were destined to lose was at its peak. I arrived not long after the famed Tet Offensive and spent part of 1969 flying along on bombing missions with the Colorado Air

National Guard, accompanying draftee "grunts" on jungle combat patrols, visiting remote firebases, and watching the 25th Infantry Division completely destroy, for no good reason — except that this was war — a group of pastoral hamlets, home to several thousand peasant farmers, west of Cu Chi. I visited my friend Ollie Davidson, who would later join the board of SELF, at his compound in Trang Bang where he was in charge of "civil operations" for USAID. There I was cajoled into joining a night patrol of the local anti-Vietcong militia and given a captured AK-47 to carry, along with my camera, just in case.

Back in Saigon at one of *Time-Life* photographer Tim Page's legendary substance-abusing parties at his Tu Do Street apartment, Sean Flynn (son of actor Errol Flynn) asked me to ride with him to Cambodia on his motorbike to see the war from that side. I nearly went, but thought better of it. The war-loving adventurer-photographer later made the trip with Dana Stone, a CBS reporter, and the pair were never seen alive again.

But all that was so, so long ago, and I was finally getting back to Vietnam nearly 25 years after those misspent days of rage and glory — 35 years ago as I write this.

SELF was in Vietnam, as I explained in Chapter 3, because a third of the population, six million families, most of them rural, had no electricity, and the Vietnam Women's Union (VWU) had agreed to do a joint project with SELF, and later SELCO, to electrify their rural members with solar power. Madame My Hoa, president of the VWU, introduced me to Madame Phuong, a powerhouse who managed huge social-development projects for the VWU like a Prussian general. She was assisted by a shy younger woman, Madame Pham Hanh Sam, who eventually took over the projects in Vietnam when Madame Phuong retired. (Retirement is mandatory at 55 — life expectancy is much shorter in the Two Thirds World.) While both women spoke some English and could read it, they were nonetheless always accompanied by capable young female interpreters who spoke English perfectly. (Note that the Vietnamese still use the French title "Madame" and its abbreviation, Mme — the country was originally a French colony.)

Mme My Hoa arranged for me to go with Mme Sam, local VWU officials, and an interpreter to visit a hamlet near Ho Chi Minh's birthplace in
Nge Anh Province, about 200 kilometers south of Hanoi. The provincial
capital, Vinh, had been flattened by US bombing and completely rebuilt by
East Germans, rendering it the ugliest city in Vietnam. We negotiated dirt
tracks in our Chinese army jeep, traveling 37 kilometers west of Vinh, past
the last electric power lines, and on through the low jungle scrub to the self-
contained hamlet of 1,280 households. The People's Committee of Thinh
Thanh had set out orange sodas and cookies in the community center, which
I had promised SELF would electrify to showcase solar PV in the province.
I asked if they had ever met an American before.

"We have not seen an American since one came down from the sky with
a parachute during the American War," said Mr. Dung, the thin, angular,
People's Committee chairman. "We took him to the authorities." The chairman seemed not to wish to open this embarrassing chapter of our joint
history, however, and I let the matter drop and smiled. "Let bygones by
bygones," he said.

Turning to the matter at hand — electricity — Mr. Khoan, the district
People's Committee chairman, told me that Thinh Thanh would not get
electricity for ten years. The commune officials thought maybe sooner, but
Mr. Khoan said, "When I say we won't have electricity for ten years, that
means that!" He then informed everyone that it would cost the commune the
equivalent of $1.2 million dollars to bring it here. There were no more sub-
sidies since *doi moi* ("open reform") began in 1986.

Mr. Dung said, "This commune has many innovative programs. No
more collectivization. We are allocating agricultural and forestry land to families. We will take care of the solar project. People will pay. It is not like before
when things were free. We want to bring the light to improve people's lives."

For less than half a million dollars we could have electrified everyone
with solar, but we didn't have that kind of money. With the first grant funds
I had previously raised from the Rockefeller Brother's Fund we could electrify 30 households, plus the community center (for which the hamlet would

have to pay the Women's Union). Thirty families had already signed up, and the local VWU official, Mme Lan, who was exactly the same age as me, was very excited about the program and agreed to collect the monthly installment payments from the families, as well as from the commune government for the larger system that would power lighting, community television, and the police radio at the community center.

The average annual income here was $400 to $600 per household. Our smallest SHS cost us about $330, which SELF passed along without markup, but that was still a lot of money, even on a four-year low-interest credit term. Yet the people eagerly paid their deposits to get on the list. Besides growing rice and vegetables, they earned income by contract sewing. Apparel and cotton gloves were trucked in, already cut to pattern; sewn on treadle machines by men and women in various households; then shipped back out. But there was no light to sew by in the evenings, when most of this extra work was done. Our lights, we later learned, had a big impact on the community. I was told 750 families already had battery-operated radio-cassette players, and 40 families had black-and-white TV sets, which they hooked up to cheap batteries that they hauled to town once a week for charging. While I was there, groups were clustered around several TVs watching World Cup soccer. The signal was being pulled directly from the satellite by simple antennas. No one looked up from the soccer tournament to notice the first American to ever visit their community.

Down the spine of Vietnam, from the huge Hoa Binh hydropower project in the north, all the way to Ho Chi Minh City, ran a huge 500 kV transmission line stretched across 150-foot towers spaced thousands of feet apart. With sufficient hydro resources in the north, and growing power demands in the booming south, this was an internal Vietnamese solution to a Vietnamese problem: they had proudly built the line themselves. The amazing thing about it, however, was that the huge high-voltage line rose and drooped, from tower to tower, across a thousand miles of inhabited land. Tens of thousands of people lived within sight of, even directly underneath, these soaring electric wires but had no access to electricity. However, everyone to whom I pointed this out was smart enough to know that it was too costly to step down 500 kV for residential service, and if communities along the way tapped this power, there would be none left for Ho Chi Minh City. Our jeep passed back and forth under this power line as we drove to and from Vinh and other communes.

As Sam and Phuong and I rode back to Vinh with the members of the local Women's Union, it seemed that bygones were not gone. The VWU's Mme Lan told me that she had just missed direct hits by US bombers three times, and that 64 people in her village had been killed. American authorities said the United States was just targeting the nearby bridge. As a journalist many years earlier, I had interviewed fighter pilot POWs who had been assigned to hit the particular bridge north of Vinh she was talking about, which we had driven over. The pilots said they were never able to knock it out. So much for precision bombing.

Sam lightened the moment with her own observation. "When we were children, it was very exciting for us when the planes came over — we saw many, many. We liked to hear the 'boom, boom'."

As I flew down to Ho Chi Minh City, I saw more reminders of the war: the row upon row of hardened concrete revetments at Tan Son Nhut Airport that once housed our Phantom fighter jets. They still stood, good as new. No one could knock them down; they had been built by Kellogg, Brown & Root, now part of Halliburton. Ho Chi Minh City mostly looked the same, with rivers of motorbikes flowing up and down the tree-lined avenues. Many were driven by women with flowing *ao dais*, the formal pant-dress, and many perilously carried whole families, stacks of produce, or cartons of appliances through the nightmare traffic.

We were to meet Mme My Hoa in the delta at the provincial river town of Tra Vinh. We boarded boats, hers a naval patrol craft and ours a small, wooden in-board, and headed through the Mekong's maze of islands and canals to reach the hamlet of Phu Dong. The women on board, all members of the national, regional, district, and local VWU, plus Sam and Phuong, broke out fruit and soft drinks on the deck and had a party as we navigated the kind of narrow waterways that had been memorialized in the film *Apocalypse Now*.

Mme My Hoa, in fact, had led a local Vietcong unit in these low-lying islands and mangrove swamps; this was her home, although she was now based in Hanoi. She had been captured during the war, I learned, although she never mentioned it, and spent many years in prison, including time in the notorious "tiger cages" of Con Son Island, the prison complex run by the South Vietnamese government. She was welcomed at Phu Dong as a returning hero. Phu Dong had been selected as another site for our solar project because it was a "revolutionary area." This meant, unfortunately, that the local "war hero" families expected to get their solar systems for free, as an entitlement for their service in the American War.

Hundreds of village children crowded around, staring wide-eyed at the foreign visitor. Mme My Hoa said, "See all the people that have come to have a look at you!" The local VWU representatives told her that they had already signed up 100 families for the SELF-VWU solar program and had collected the initial deposits. "They are saving for the light," the local WVU official told me. "Many people are suffering from living and dying in darkness."

I certainly never imagined, when I flew over the green expanse of the delta in US helicopter gunships in 1969, that one day I'd be down here in this dense, stiflingly hot undergrowth selling solarelectric systems to peasant farmers and former fighters for the Vietcong.

Phu Dong and its neighboring hamlet, Phu Tan, were home to some 3,000 families who grew rice, fished, and raised shrimp in large excavated shrimp ponds. Because they lived on delta islands separated from the mainland by miles of water — the Mekong's fingers are amazingly wide here — they had no electricity and no prospects of getting any. Perfect for solar.

The next time I returned to Phu Dong and Phu Tan, a large solar array of huge, glass, amorphous-silicon modules stood on a cement-anchored frame in front of the community center, VWU offices, and clinic. The oversize 200 kW modules had been donated to SELF by a US solar company, and I'd arranged for them to be shipped to Ho Chi Minh City with one of our container loads. They produced more power than the center knew what to do with. "Please don't send us any more of those," Mme Phuong said. "We couldn't even lift them!" But they had transported them by boat and installed them here, along with a hundred 20 W SHSs installed in households scattered through this very remote community. These distant hamlets could boast of being the genesis of solar rural electrification in Vietnam, and the Women's Union was very proud of its ability to undertake a technical project of this kind.

"I was very worried about this new technology, and worried that people couldn't pay back," Mme Hanh, president of the VWU in Tran Vinh, told me. "But the solar works, we overcame our difficulties with it, and now the light contributes to the people's material life, their literacy, and gives opportunity for income generation and expanded knowledge." I couldn't have expressed it better myself. "Seeing is believing," she added, "as we say in Vietnam." We say it too, I told her.

To assist the women, I sent an experienced PV technician to Vietnam: Marlene Brown, a former solar-system installer from New Mexico, was then working as a technical expert at Sandia National Laboratories (which helped finance some PV systems in Vietnam long before the United States had reestablished diplomatic relations). Marlene, who also worked for SELF in the Solomon Islands, had more enthusiasm than just about anyone I had ever met, and she also possessed the technical skills required to train, with the help of interpreters and Mme Sam, dozens of local women to install and maintain SHS. She was fearless, tireless, funny, and always the diplomat who could overcome any confused situation with humor and understanding. With the help of Solarlab, a technical outfit in Ho Chi Minh City that manufactured some of the electronic components we used, Marlene and her all-woman crews could install 10 SHSs a day, nearly 100 a week, in conditions that would drive any normal electrical contractor from the West to despair. Marlene patiently taught every family how to use their SHS, and she

Marlene Brown

PV technician-trainer Marlene Brown raising an amorphous solar module with Vietnam Women's Union members in the Mekong Delta.

photographed every solar house SELF electrified. She spent many months in Vietnam during several visits and became close friends with her contemporaries in the VWU, who adored her.

The only thing the women couldn't — or wouldn't — do was climb onto roofs to mount the solar modules. "Do you mind if we ask the boys to do it?" Sam asked me shyly one day. "They have offered to help us. Women don't climb on roofs." But they did manual work, such as carrying heavy batteries to the boats for loading, then unloading them on the islands and carrying them on poles to the individual houses. Transporting hundreds of solar modules to these remote locations was equally burdensome for the women.

In Tra Vinh Province, farther south, we launched projects in two more Mekong island communes, Hoa Minh and Long Hoa, home of some 6,000 rice growers, shrimp farmers, and fishermen. To visit these houses we walked along narrow dikes separating the paddies, crossed single-log "monkey bridges" over the irrigation canals, and traveled muddy paths through the thick vegetation. Every once in a while a simple thatched house with a tall pole supporting a solar module would appear, and the family would welcome me inside to see their lights, their TV, and often their stereo, some of them equipped with a microphone to be used as a karaoke system, all powered by our solar PV. In one house, an enormous python rested in his cage with a duck sitting on top of him, unawares. The owner said, "When he gets hungry, the duck is lunch." He would eventually sell the python for its meat and earn funds to put toward his solar system.

The president of the Tra Vinh Women's Union, Mme Khanh, never smiled nor showed the same enthusiasm Sam and Phuong had for our project, and I learned why: Her husband had been killed by US forces, and I was the first American she'd ever met. It was a testament to the humanity of the Vietnamese, and their attitude of national forgiveness for Americans and our war, that they would even talk to their former enemy, an enemy who had come from 10,000 miles away to destroy their land and kill close to a million of them. (Their humanity stood in stark relief to the suicidal barbarity of the Islamists in the Middle East and Iraq, and the maniacal hatred of the 9/11 terrorists.) Let bygones be bygones. Right, but not so easy.

SELF and the VWU launched six separate solar projects in Vietnam, two in the north and four in the south. One of the northern villages was near Hoa Binh and the country's largest hydroelectric dam, but the people had no power. They earned money to pay for their SHSs by raising puppies for the dog restaurants in Hanoi. I was glad this was one community I never visited; I only saw the photos. (I didn't tell my wife, the executive vice president of the Humane Society of the United States!) We installed several hundred systems in Tien Giang Province, also in the delta, where I suffered through embarrassing drunken noontime feasts with the local People's Committee and Communist Party chiefs. Sam and Phuong were appalled, but there was little they could do when a local party functionary decided a visit by a foreign guest was an occasion to drink himself and his buddies under the table on the government's tab. The food, however, was terrific, except for the duck eggs with fully formed, unhatched ducklings inside — very good with the fermented fish sauce called *nuoc mam*, I was told.

In Hanoi we invited the government bureaucrats from the ministries of industry, finance, planning and investment, science and technology, agriculture and rural development, and representatives from Electricity of Vietnam (EVN), the World Bank, the IFC, and UNDP to an all-day "national solar-electrification seminar" at the top-floor conference center of the VWU's headquarters building. Women came from all over Vietnam, from all levels of the VWU, dressed in their finest *ao dais*; they were there to learn about solar or to provide testimonials on how solar electricity had changed lives in their homes and communities. Peter Riggs from the Rockefeller Brothers Fund attended, at my invitation (see the photo in Chapter 5). Sam, Phuong, Peter, and I sat on the dais next to a large plaster bust of Ho Chi Minh; simultaneous translation through headsets allowed speakers to use Vietnamese or English. All day long we heard how electric light had transformed lives, how the local women had struggled to implement the financial and technical aspects of the project, and how proud they were of their ultimate success, electrifying some 1,000 families with no government support.

We shipped containers of US-made solar modules and batteries to Vietnam, duty free, and Mme Sam and her patient staffers managed to get the shipments through the corrupt customs (eight custom officers were executed for corruption while I was there) and delivered to their storage sheds on the grounds of their southern headquarters in Ho Chi Minh City. The VWU's offices there were located in the former villa of General Westmoreland, who commanded the US war in the 1960s.

It was in General Westmoreland's old living room that I attended a private dinner with the woman in charge of the southern branch of the VWU, Mme Thang. In her mid-40s, she was stunningly attractive with a warm smile and a no-nonsense manner. She wore a colorful silk dress, not an *ao dai*, and shimmering jewelry. Women's Union staff served us from the villa's kitchen as we talked about the solar project and the good work Marlene Brown was doing. As dinner progressed, we reminisced about the war years.

"I spent six years in prison," she offered, "from 1969 until liberation."

"For what?" I inquired.

"I was a member of the National Liberation Front," she said, smiling.

She hesitated to say more, but I inquired. My own antiwar credentials helped me here — we both knew some of the same people in the US antiwar movement, like Cora Weiss and John McCauliff. I told her a close friend had started Vietnam Veterans Against the War.

So she put down her chopsticks and said, "I tried to assassinate one of the top ranking ARVN [Army of the Republic of Vietnam] generals ... in his bed."

"Oh," I murmured.

"I had pretended that I would be his mistress so I had access to the house, and one night I went to his bedroom, lifted the mosquito net, and pointed a big pistol at his head ... but I couldn't do it. Maybe I could have, but before I could decide, he woke up, grabbed me, and I was arrested."

As a lovely young woman in 1969, she was a celebrity at her trial in Saigon, where a judge sentenced her to life in prison. When the verdict was read, she smiled what became a very famous smile, photographed by AP and sent around the world by newswire. The judge asked her, "Why are you smiling? I just sentenced you to life in prison. That is the smile of a crazy person."

"No," the young Mme Thang answered, "it is the smile of victory."

She was let out of prison when the National Liberation Front and North Vietnamese Army overran Saigon in April 1975. Today she is the minister of tourism. The "smile of victory" photo still hangs in the Vietnam Women's Museum, which chronicles the role of women in Vietnam since the Trung Sisters drove out the Chinese invaders in AD 40.

Shawn Luong, a senior engineer with the Los Angeles Department of Water and Power, contacted me one day in Washington after reading an article I wrote in *Solar Today*, a trade magazine.

"This is the right technology for Vietnam," he said. "I'm familiar with PV. I've installed many mountaintop communications towers in California for DWP, powered by solar. It would be good to use PV in Vietnam where rural people have no electricity."

I said we already were.

Shawn flew to DC and offered to help. He also wanted to set up a joint venture with SELF in Vietnam, and I said we were considering forming a company, SELCO, and that we had already done it in India. Shawn (not a regular Vietnamese name, as one might suspect) was one of the "boat people." He and his parents escaped from Vietnam in 1979, four years after "liberation." His father owned lumber mills, textile factories, and farms near Saigon, which were confiscated by the new government. Shawn, who spoke French as well as fluent English, had attended the Lycée Madame Curie; he was 21 when they set off on a small boat, survived pirates in the South China Sea, and washed up in Thailand where they were taken to one of the big UN refugee camps set up for the boat people. Being a natural hustler, it wasn't long before he managed to get his family, his future wife, his brother, and himself to Los Angeles, arriving with a total of $200 in their pockets. The new communist government had confiscated all their savings.

Shawn came from the Chinese community in Ho Chi Minh City, which had been there for several hundred years, blending in but retaining the language. He told me, "We Chinese are the Jews of the orient." Chinese, Jewish, Vietnamese, or whatever, and despite the fact that he was a Republican supporter of Ronald Reagan, I found Shawn immensely likeable, honest, and hardworking, although extremely cocky (to use his own word). I figured I could not find a better partner with whom to launch SELCO-Vietnam. I certainly couldn't do it myself; I didn't speak the language. He had been running a trading operation on the side and had gone back in 1990 to meet the rulers in Hanoi who had driven him and his family out of the country 15 years earlier — a very different approach to reconciliation than is practiced by Miami's Cuban exiles. He wanted to do business with the Democratic Socialist Republic of Vietnam. With a graduate degree from the University of Southern California and a successful 12-year career at LADWP, he was ready for something new. We agreed his trading company offices in Ho Chi Minh City would be the new SELCO-Vietnam headquarters, and he set about securing the first license for a foreign-owned solar company to do business in Vietnam.

In December 1997 we received a nice, blue-covered bound book containing our license to sell solar devices and equipment in Vietnam through SELCO-Vietnam Company Ltd. We had very big plans.

Shawn quickly made friends with the VWU representatives, including Phuong and Sam, and I tried to make friends during my visits — now accompanied by Shawn — with the local World Bank staff, the managers at the new office of the IFC, and the officials at the Ministry of Industry and Electricity of Vietnam. We had the biggest development organization in the country behind us, with connections in the politburo and a track record of hundreds of installed SHSs, so why shouldn't we think big? Communist governments like big plans, usually in five-year increments, so we set out to develop a five-year solar-electrification plan with the support of the VWU. It would be the largest commercial solar project in the world. Or so we hoped.

It was in Vietnam that I was to learn all the lessons I would ever need to learn in business, politics, economic development, corporate management, human relations, cultural barriers, and global finance. I'd already learned a lot about international affairs and war from Vietnam 25 years earlier; how much more was there to know? Lots.

Shawn got to work, mounted a neon SELCO sign on his five-story glass office building on Nyugen Thi Minh Kai Street, electrified the SELCO office floors with solar PV to assure power for the lights and computers during power cuts, and we opened for business. Shawn commuted regularly to

Los Angeles, where his family continued to reside, so he put Tran Thanh Danh, a longtime family associate, in charge. Danh (last names are first names in Vietnam) was a self-taught electrical engineer and an energetic worker with a beatific face and infinite patience.

Sam and Phuong identified remote villages where the local VWU leaders verified there was little probability of electricity arriving, and where people expressed interest in acquiring a solarelectric system. We expanded operations in areas where SELF had already launched projects, and we opened solar-powered solar service centers. We moved into new provinces — Phuoc Vinh, Dak Lac, Ca Mau, Binh Thuan, and even Phu Quoc Island off the Cambodian coast. We opened

Satisfied customer in Binh Phuoc Province, Vietnam, with Tran Danh (right), SELCO-Vietnam managing director.

service centers as far north as Ban Me Thuot in the central highlands. Shawn was ambitiously spreading us pretty thin, but we counted on our unique and exclusive relationship with the VWU to bring in the customers.

I transferred from Citibank Washington to Citibank Ho Chi Minh City a large chunk of money from the parent company to

SELCO-Vietnam, the minimum capitalization required by our Vietnam business license. We then borrowed another large sum from the IFC in Washington, the same people who were driving us nuts in India with their Photovoltaic Market Transformation Initiative. We never seemed to learn. However, they were more amenable to lending money to finance solar rural electrification in Vietnam, thanks to staffers at the IFC who truly understood development banking. In the end, it all comes down to people, in this case one Doug Salloum at the IFC. Doug had visited one of our service centers in Tra Vinh during a power outage, and the only light on the street — even his hotel was dark — emanated from the solar-electrified SELCO-Vietnam office.

To close the deal we brought Phuong and Sam, their smart young interpreter, and a senior VWU official to Washington via LA, where Shawn took them to Disneyland and Hollywood. He also took them to Mexico to visit the assembly plant of Detroit-based United Solar Systems, which was then providing our solar modules. We had dreams of opening a PV-module assembly plant in Vietnam that would be staffed by women workers. Phuong, Sam, and their companions were suffering from jet lag — this was their first trip to America — and I'm not sure they knew exactly what they were looking at. Since they did not have multiple entry visas, they were nearly denied readmittance to the United States. Shawn talked Immigration into letting them come back. Then he took them to Disneyland, in his big black Mercedes, and on to see his suburban split-level home on a beautifully landscaped hillside overlooking the Pomona Valley, where they asked him pointedly, "Why are you leaving all this to come back to Vietnam to help us?"

I'd also wondered.

"My priest at our Buddhist Temple in Garden Grove told me I had to do it," Shawn said. "My karma required it." People had often asked me the same thing — "Why do you do this?" — but I never had such a clear answer. I wasn't doing it to get rich, that was for sure. I discovered the hard way that this rocky, risk-laden path was not the road to wealth. But if wealth were what I sought, I could have found a far easier way to acquire it. So Shawn and I were somewhat on the same page in this regard, except that he also ran a trading company, sold real estate, brokered mortgages, and invested heavily in Cisco Systems, all the while holding down a senior job at LADWP until he joined SELCO-Vietnam.

In DC I arranged a formal lunch for the VWU delegation with a World Bank vice president or two at one of the executive dining rooms in the Bank's gleaming new building. After lunch I took them to meet the officials at the IFC, down the street on Pennsylvania Avenue, who were considering our loan proposal. The IFC thought the VWU should borrow the money, since the solar loans were going to be made to Women's Union members, but the women were smarter than that; they didn't want the liability. They would take grant money and spend it wisely and honestly, but they rightly feared debt. So SELCO took the $750,000 loan, along with all the risk and all the liability. Our company's future now depended on collecting installment loan payments from peasant farmers in Vietnam.

To introduce our new company to the VWU, I participated in meetings in Can Tho, My Tho, Tra Vinh, Ho Chi Minh City, and Hanoi with the VWU. Mme Phuong fully understood the market economy and the need to do this commercially and not simply as a subsidized, charitable, social program, which it had been thus far. She urged the VWU officials to support, in this case, "business integrated with social development." The VWU was already the largest commercial tour operator in the country, and it owned the Apple Computer franchise. The officials understood business.

And so in 1999 we proposed a 12,000-house solar project as the first phase in a plan over the coming decade to reach a million homes. The vice minister of industry, in charge of electricity for Vietnam, told me in a formal conference room at the ministry, with its plush chairs, lace doilies, lacquered coffee tables, and porcelain tea sets, that "we can never electrify these people," referring to the last two million families who would still remain in the dark 20 years hence. "We must use solar," he emphasized, pounding his chair arm.

We also gained the tacit support of EVN, whose dour, skeptical engineers had come around to our view that solar PV provided the cheapest and fastest way to electrify communities remaining far beyond the grid. For now, they said, they would leave it to the VWU, since they had their hands full operating and maintaining the conventional power and distribution system. They asked us to keep them informed of our progress.

Meanwhile, Shawn set up a large warehouse on Ho Chi Minh City's outskirts and we began manufacturing our own charge controllers and light fixtures. We imported German compact fluorescent lamp technology from China and began assembling energy-efficient AC and DC bulbs in our workshop. Soon, half the hotels in the city were illuminated by SELCO-Vietnam's low-cost, high-quality, compact fluorescent lights, and our DC models brightened the homes of our growing base of rural customers. I was constantly amazed at the technical ability of SELCO-Vietnam's employees; they could fabricate just about anything, from PV-powered railroad signal lights to lamps with clusters of light-emitting diodes (LEDs), which use almost no power, to solar streetlights and DC fans.

On the marketing side, an enormous SELCO billboard, illuminated by 200 watts of solar power and our biggest compact fluorescent lights, towered over the central plaza at one of the Mekong River ferry crossings. It proclaimed in Vietnamese, "Solar Light Brings Joy To The Home!" The VWU held training sessions for "motivators" all over the south, and we set up booths at agricultural trade fairs, where we displayed our solar product line to the amazed rural and urban clientele. The VWU hostesses sported SELCO baseball caps; our technicians wore SELCO polo shirts.

To finance our customers, Shawn and senior officials of the VWU convinced the Vietnam Bank for Agriculture and Rural Development (VBARD) in Hanoi to sign an agreement creating a program whereby certain of their 2,500 branches would make solar loans, collateralized by our funds from the IFC. VWU members would help the local VBARD branches qualify borrowers, and for a small fee, on top of their sales commission, they would also collect the monthly installments and turn the funds over to the bank. This was probably the most sophisticated solar finance program ever put together anywhere — on paper. And it worked — for a while.

I visited some of our customers in Binh Phuoc Province, out near the Cambodian border, just beyond the so-called Iron Triangle, where the United States had denuded over a thousand square kilometers of forest cover with toxic dioxins (Agent Orange) and removed the entire population to fortified "strategic hamlets" surrounded by barbed wire. With no people it was presumed the Vietcong couldn't fight a "people's war." The people became refugees in their own land. This was the bright idea of one Herman Kahn of the right-wing Hudson Institute, which would later advise the US government on Iraq.

Now this fractured landscape, with its new growth, had been opened to recolonization, and families from the crowded Red River delta in the north had been offered land to farm in the Iron Triangle since the Americans had so kindly removed the trees. These were SELCO's customers, growing black pepper on neat plots, nursing the land back to health.

In one two-story wooden house half covered in blooming bougainvillea, with mounds of peppercorns drying in the courtyard, Danh introduced me to a middle-aged peasant woman with a wide smile. She proudly showed me her house, careful not to disturb her aged father sleeping in the living room. She switched the lights on and off in each room and said she had more lights upstairs where her son studied most of the night. I asked how the system was working. She had bought, with the help of VBARD and the VWU, a 75 Wp system, our largest.

"Oh fine," she said. "But when my son studies, sometimes we don't have enough power for our color television." She switched on the set, which seemed to be working, and I marveled that she could afford a color TV. (Our technicians rewired color TVs to operate on direct current.)

"What do you do then?" I asked. "Tell your son not to study anymore? Or stop watching TV?"

She promptly showed me her answer, opening a wooden cabinet on which the color TV stood. Inside was another television. "Oh, we watch our black-and-white TV. It uses less power." She had figured out what we called "power management." If you used solar power wisely, there was usually enough for all occasions during most of the night.

In other nearby houses I found pepper farmers with excited children watching taped programs on their solar-powered VCR in black and white, since they couldn't afford a big enough PV system to run lights, a VCR, *and* a color TV. Most of the families had also bought a SELCO fan to move the

humid air through the main room on a hot day. All through this revitalized war zone, 100 kilometers northwest of Ho Chi Minh City, SELCO lights were keeping the dark tropic night at bay.

A pepper farmer in Vietnam shows off dual TVs, one color and one black-and-white, which she uses to save power when her son needs solar lighting to study.

Vietnam is a long, skinny, coastal country, with its capital, Hanoi, about 1,200 kilometers north of its commercial center, Ho Chi Minh City. Vietnam Airlines' air shuttle connected the two with Airbus 320s and Boeing 767s. (I had earlier flown on one of VA's last Russian Tupolevs before it crashed, as they all seem to do eventually.) Nothing happens without government approval at the highest levels, which can only take place in Hanoi, so Shawn found himself spending a great deal of time there. He chased after the military, the police, the forestry department, the ministry of ethnic and tribal peoples, EVN, and he wined and dined a friend in the prime minister's office, who promised to help us launch a huge national solar-electrification program with the PM's backing. First, however, we had to sell the idea to the national assembly.

Shawn and the VWU worked together to put a SELCO exhibit right smack in the middle of the grounds of the national assembly during its annual meeting. It was a small, solar-powered, pre-fab house, with recessed solar lights and PV modules integrated into the roof. It was a sensation, visited by the media and hundreds of members of the assembly. No private company had ever been allowed to display its goods or products on the grounds of the national assembly. Shawn wore his best pinstriped suit, Mme Sam her best silk *ao dai*. Then Nong Duc Manh, chairman of the assembly, came by for a visit, and Shawn and Sam explained to him how solar electricity worked. He was fascinated. Shawn posed for photos with him, and shortly thereafter Mr. Manh was named secretary general of the Communist Party, the country's highest leadership post. Boy, were we connected now! No capitalist enterprise in Vietnam was complete without the blessing of the country's top communist. So it goes.

Nonetheless, doing business in Vietnam was not easy, we learned the hard way. SELCO managed to install only 2,200 SHSs in its first few years, and the company found itself financing many of these customers itself. VBARD approved fewer and fewer loans because it could not lend to families that already had an outstanding agricultural loan with the local VBARD outlet. Or because the bank employees were lazy and just didn't feel like it.

Or because people were afraid to go to the local bank, preferring to keep their money under their mattresses — there is *no* history of, or experience with, credit of any kind in Vietnam we were to learn later instead of sooner. But people wanted electricity, and the VWU vouched for the customers and agreed to collect the payments. So we extended company credit, more and more credit to more and more families, scattered farther and farther away, all over southern Vietnam.

We prepared detailed loan documents in Vietnamese and English, which were signed by the farmer, the local VWU rep, the local People's Committee (like having the sheriff co-sign your auto loan), and our salesperson (and VBARD if it was a bank loan). We planned to refinance these loans one way or another; they were scanned and e-mailed back to Washington, where one day I showed several hundred of them to Mildred Callear, the vice president and treasurer of the US Overseas Private Investment Corporation (OPIC), who had come by the SELCO office in Chevy Chase.

Ms. Callear said, "These farmers are probably a better credit risk than some of the big power projects we finance." How right she was: our farmers are still paying their installments, while Enron, for example, stiffed OPIC and US taxpayers for the $700 million it owed on its power plant in India. Unfortunately, even though OPIC's priorities during the Clinton era were to finance women, renewable energy, and projects in Vietnam, in that order, SELCO's balance sheet didn't qualify us for an OPIC loan to refinance our farmers in Vietnam, so SELCO was stuck carrying these receivables itself (no company can finance the sale of its own products). Perhaps we needed Enron's accountants to fiddle with our balance sheet. Maybe we should have been using Arthur Anderson instead of Ernst & Young; Anderson knew how to produce a fraudulently bankable financial statement — good enough for OPIC — even if it did eventually bring down their house and Enron's. At least SELCO is still in business.

SELCO-Vietnam moved into new quarters at the beginning of the new century: a five-story building of our own on Tran Hung Dao Boulevard. The neon SELCO sunburst logo was visible from a mile away atop the building's roof. A clean, modern showroom graced the ground floor, open to the side-walk, and sometimes people walked in and ordered a PV system for their urban home as a backup against annoying power outages. Government officials came to look around and study the photos of our rural solar installations in hard-to-get-to places where they were never likely to tread. Our salespeople set off on their motorbikes every day, and our new Mercedes van carrying a technical team delivered SHSs in batches of eight or ten to rural communities within a day's drive of Ho Chi Minh City.

Up in Hanoi, Shawn and I met with the US ambassador, Pete Peterson, a former POW and ex-congressman. We told him what we were up to and our big plans to electrify the unelectrified with help from the World Bank, the IFC, the government of Vietnam, the US government, the Asian Development Bank, and God knows who else, and he offered all the support he could through the various US agencies now working in-country. Pete, like Senator John McCain, had spent years in the "Hanoi Hilton," the old French prison in central Hanoi, and now he was back — as our ambassador! At least one American was capable of forgiveness. In Hanoi's B-52 Museum (that's what it's called), featuring the exploits of the North Vietnamese air force, I saw Pete's fighter helmet on display, next to a piece of the wreckage from the fuselage of his F-4, his stenciled name still visible beneath what remained of the cockpit frame. Pete's story is an inspiration to all of us who lived through those war years; he retains no bitterness toward Vietnam or his own government. He even married a Vietnamese woman. Thanks to Pete and senators John McCain and John Kerry, the United States finally normalized relations with Vietnam in 2001. President Clinton paid a visit to the country, and Hanoi ratified a trade pact with the United States shortly thereafter.

This should have made everything easier for us, but it had little effect.

James Wolfensohn, president of the World Bank, came to Vietnam for a look-see, as Vietnam was now the flavor of the month among international donors. I had met him earlier in DC, after his big PV powwow, and invited him to see one of our solar communities in Vietnam. His money was help-ing to finance them, through the IFC. In Ho Chi Minh City, Wolfensohn met Shawn, who gave him a SELCO hat, which he wore the rest of the day to protect his head from the equatorial sun. Another photo-op.

The World Bank continued to send consultants to study SELCO-Vietnam and VWU solar projects and proposed all manner of financing through its renewable energy master plan for the country. But in the end, the conventional power guys won out, and the first $50 million International

SELCO

Development Agency loan (at the lowest interest rate offered by the Bank — 1.5 percent) for rural electrification didn't contain one dollar for solar energy. Instead, the Bank provided heavily subsidized funds for unsustainable rural electrification that would never

SELCO-Vietnam co-founder Shawn Luong (left) presents a SELCO hat to World Bank president James Wolfensohn while IFC country director looks on.

pay for itself, and the word got out that even the remotest regions would soon get electricity. Some did. We could compete with no electricity, but we couldn't compete with nearly free electricity, heavily subsidized by the World Bank — which would never subsidize solar power to the same degree.

SELCO-Vietnam struggled along, setting an example of good corporate citizenship. But our alliance with the VWU was getting wobbly; our business was hampered by the VWU's bureaucracy and institutional inability to act as our agents effectively. It was a social service organization, and we were a business, and, alas, never the twain shall meet — really. VWU members preferred providing low-cost, partly subsidized SHSs under the SELF charity model, even while they understood our business model and tried hard to support it. Clearly, collecting money for a revolving solar fund they managed was one thing, but collecting deposits and monthly installments for a private company, or for a bank like VBARD, which was too lazy to collect its own loans, was a different kettle of *nuoc mam*. More and more this unusual, commercially financed, social-development, public-private partnership made less and less sense, and SELCO opted to go it alone. I think the VWU was grateful. Nevertheless, in 2001, Mme Sam was flown — deservedly — to Vienna to receive the 2002 Energy Globe Award for bringing solar electricity to women in Vietnam. The European organizers who bestowed the award at a black-tie gala attended by half of Austrian high society never credited SELF or SELCO — not so much as a mention.

For our ten-year effort fighting the Vietnam solar war, SELCO received a prestigious award of its own in 2001. I received a call from Undersecretary of State Alan Larson to tell me we had been nominated in the small and medium business category for an Award for Corporate Excellence, bestowed annually by the US State Department. A letter followed to say we were, indeed, the award recipient. It was to be presented in December by Secretary of State Colin Powell at a ceremony at the Dean Acheson Auditorium at the

State Department. Furthermore, there would be a live video hookup from the US consulate in Ho Chi Minh City, where SELCO-Vietnam's staff would simultaneously receive a copy of the award from the new US ambassador to Vietnam, Raymond Burghardt. This was pretty exciting for our little company, to say the least.

Shawn had turned over the reins in Ho Chi Minh City to Canh Tran, a public administration graduate of the Kennedy School at Harvard and a war refugee who had flown out of Saigon in the back seat of a South Vietnamese Air Force fighter jet as the North Vietnamese made their way into the city in 1975. Like so many Vietnamese refugees, he had achieved the American dream, with a suburban house and two kids in college, and like so many he wanted to return to his homeland and bring his skills with him. SELCO named Canh managing director of SELCO-Vietnam in 2001.

Canh and Danh put together their half of the State Department award ceremony in Ho Chi Minh City. Danh built a model thatched house inside the consulate's studio, complete with PV module and lights, and mounted SELCO photos of rural installations for a backdrop. They invited top officials from the VWU and the People's Committee of Ho Chi Minh City, along with exemplary SELCO-Vietnam workers.

The ceremony was postponed until January 16, 2002, to accommodate Secretary Powell's schedule. He confirmed the new date, but that morning he found himself on an unscheduled flight overseas for a meeting on the war in Afghanistan at the president's behest. State's number two, Richard Armitage, was set to do the honors, but he was called over to the White House at the last minute, so Assistant Secretary Mark Grossman delivered the speech and presented the granite and crystal award on which was engraved SOLAR ELECTRIC LIGHT COMPANY: For Outstanding Corporate Citizenship, Innovation, and Exemplary International Business Practices in Vietnam.

I gave my speech before the audience of retired diplomats, desk officers, foreign embassy officials, the Vietnamese ambassador, Shawn Luong, SELCO staffers, and my wife, Patti. SELCO videos from Vietnam were shown on a giant video screen and then the picture went live to Ho Chi Minh City as the

American ambassador made his remarks, which were followed by Canh's expressions of gratitude for this prestigious recognition.

Canh said, "I would like to thank my staff here in Vietnam — here in Ho Chi Minh City and at all of the branch offices — for their hard work, day in and day out. I am so proud of what they and their colleagues have done, not only for SELCO, but for so many families who now have a chance to eat dinner together, talk together, study together, even watch TV together in their homes under their own electric lights." Behind him, Mme Sam beamed, Danh smiled, and the People's Committee representatives stood stiffly, staring into the camera.

The State Department had outdone itself with this international live satellite hookup, the elaborate ceremonies, and the reception that followed, all to honor our little company and the Ford Motor Company South Africa, the big-company recipient (which necessitated a live hookup with the US consulate in Pretoria). I felt our workers and the women had earned it.

Meanwhile, Harish Hande and Susantha Pinto from India and Sri Lanka were now making regular visits to Vietnam to help consolidate our far-flung operations and assist with overall management. Canh visited Sri Lanka and Danh visited India to compare notes on operational methods and practices. Shawn and I, and later Canh and I, continued to dream our impossible solar dream for Vietnam: to develop the largest rural solar project in the world. BP Solar was already advertising that it was undertaking the world's largest solar project, in the Philippines. But it was totally paid for by donor aid and was completely unsustainable, another solar energy giveaway. But it made a nice profit for BP. Since we were the only real player in the commercial solar energy field in Vietnam — now with some 4,000 houses electrified — we felt we had a chance to do what BP was claiming to have done in the Philippines. We were also the Siemens Solar dealer for Vietnam, and Siemens management in Munich and California were behind us. I went to Manila and got support for a massive solar initiative in Vietnam from the highest levels of the Asian Development Bank.

We lined up a half dozen provincial governments, which promised to pay half the costs of a solar subsidy themselves, recognizing that it would be cheaper to do this than attempt to distribute their costly grid power to so many remote hamlets. We had meetings, meetings, meetings with every chairman and vice chairman, minister and vice minister, director and vice director, chief and deputy chief of every agency, ministry, department, branch, and division of every level of government in Vietnam, from district to province to region to national. We joined Vietnam Airlines' frequent flyer program to log our miles shuttling between Hanoi and Ho Chi Minh City.

But it slowly became clear that in a communist state where no one is elected and there is no accountability, there is no need to make a decision if you don't have to. Officials loved meetings, but they hated making decisions, and when they did make them, they were always contingent on additional decisions being made higher up by faceless authorities who didn't like meetings and who loved to postpone their decisions. The process reminded me of the *New Yorker* cartoon of a businessman looking at his appointment calendar and speaking into the phone, saying, "No, Thursday's out. How about never? Is never good for you?" *Never* was good for the people who ran Vietnam.

What is more, our unique enterprise was caught between being seen as a social development organization and, the reality, as a strictly commercial retail venture that happened to be performing a valuable social service. "Service," however, was not a word that could be translated into Vietnamese, just as they have no word for "marketing." We got tangled up with the government because of our earlier alliance with the VWU, and also because we agreed with the VWU that to provide electricity to the very poor and unserved required government subsidies. The provincial officials offered to pony up a 40 percent subsidy, but the national government and all the foreign donors would only talk about it. Meanwhile, the World Bank arrived with $400 million to extend the grid to a thousand rural hamlets at a 90 percent subsidy. This caused our provincial chiefs and their People's Committees to reconsider. Why should they give 40 percent for solar when Hanoi was going to provide 90 percent (with World Bank funds) for grid extension? The only problem was that the World Bank program would take years and was, in the end, only going to electrify one-sixth of the six million families still in the dark.

Our grand plans suffered from dithering and procrastinating, just like Vietnam does, where everything moves with determined lethargy and purposeful tardiness. Except, of course, for the rivers of motorbikes — everyone in a hurry to get somewhere to do nothing. Despite our "innovation and exemplary ... business practices," SELCO-Vietnam finally back-burnered the big plans that threatened to burn everyone out and focused again on day-to-day business, which it still does, bringing solar power and light to those in need, those in the dark, wherever they are, providing they can afford it — without any government or donor subsidy. SELCO-Vietnam impatiently awaits the day when the people of Vietnam choose to cease their stubborn procrastination and resistance and join the modern world. That day will come when the last surviving statue of Lenin anywhere, solemnly standing in a public square, is finally pulled down.

On one of my trips to Vietnam, newsstands everywhere were selling the *Time* magazine issue with former senator Bob Kerrey on the cover, illustrating the "Ghosts of Vietnam" feature story in which Kerrey admitted that his squad had killed 21 unarmed civilians in the hamlet of Than Phong in the eastern delta 32 years earlier. Witnesses, both American and Vietnamese, said Kerrey's men had opened fire on 14 women and children no more than six to ten feet away and also slit the throat of an old man before beheading him with a knife while Kerrey himself held the man down. All this was big news at the time — a former US senator and decorated "war hero" admitting to war crimes — and was covered in the *New York Times* magazine and featured on CBS's "60 Minutes." The Vietnamese were discussing it too.

I talked about it with Danh when he got back from installing SHSs in the field. I mentioned the name of the hamlet, Than Phong in Tra Vinh Province, and asked if he had heard about the story.

"I was in that place yesterday," he said. "Than Phong. We just installed seven 40 Wp solar home systems there, in Than Phong. They all paid cash. I didn't know this story about Kerrey."

We all have our own Vietnam War. It's over if we want it.

# CHAPTER IX

# Cirque du Soleil

A s the new millennium dawned, it seemed everyone was getting into the game of delivering solar energy to developing countries. Solar "projects" were proliferating, each one accompanied by three graduate students writing PhD theses, at least five World Bank or Global Environment Facility (GEF) managers, and probably a dozen highly paid international energy consultants, economists, and "experts." It was truly becoming a circus.

We used to joke that the number of solar home-lighting systems being installed worldwide by various entrepreneurs, corporations, governments, and donor organizations was only exceeded by the number of PhDs being turned out in "energy economics," "rural development," and "social impacts" relating to the field of renewable energy and solar PV.

But the only doctorates in solar worth a damn were in solarelectric *engineering*, and American universities had turned out fewer than a dozen of those. We had hired two of them: Dr. Harish Hande and Dr. Priyantha Wijesooriya. I was always amazed to find myself hiring so many PhDs and managers with master of science degrees.

As for the cirque du soleil, it appeared in full force at a gala reception at the Swiss embassy in Washington, DC, in the summer of 2000. The Swiss government had contributed a large chunk of money to a new $30 million fund for solar energy, the Solar Development Corporation, which was one part, along with Solar Development Capital (SDC) and Solar Development Foundation (SDF), of the Solar Development Group (SDG), an outgrowth of the Pocantico Conference. Brooks Brown, the "recovering venture capitalist" I had invited to Pocantico, was initially the moving force behind SDG, which he incubated through the Environmental Enterprises Assistance Fund (EEAF), his organization in Roslyn, Virginia. This combine, well-funded by the Rockefeller Brothers Fund, the World Bank, the IFC, and USAID, was staffed by highly paid managers.

I was standing there with a lemon soda in my hand, looking around the room at almost 100 well-groomed men and women nicely turned out in dark business suits, the official Washington attire. Richard Hansen, formerly of Enersol and now of SOLUZ (see Chapter 5), was also there with a new, respectable haircut and a trim sport coat. I was encouraged, ten years after I had founded SELF and four years since launching SELCO, that now a commercial business development and investment fund was being launched to finance — exclusively — solar power for the developing world.

But as I looked around the embassy's reception room, where black-tied waiters served up canapés and drinks, I realized that only two people in this room actually had companies in place, operating in the developing world, that could use these funds: Richard Hansen and me. So who were all these high-salaried Washingtonians, working in this somewhat specialized field of solar rural development, hailing from various NGOs, the World Bank, USAID, UNDP, GEF, the IFC, and God knows where else, plus the 20 or so staff of SDG? I knew many, but not all of them, and not all of them knew who Richard or I was.

Ten years earlier I couldn't find anyone in Washington who could spell photovoltaics or who would believe that poor farmers in the Two Thirds World could actually be provided with electricity from the sun.

Now solar rural electrification was a veritable industry, which could be very lucrative indeed, provided you worked for one of the aforementioned organizations but didn't actually get your hands dirty or take risks, trying to run a business that installed solarelectric systems in rural areas of developing countries. That is the *last* thing you would want to do if you wished to pursue a career in "solar energy for development."

Anyway, despite all this inordinate overhead, I was glad that at last international, low-cost, dedicated, public funds would be available to finance SELCO's growth and underwrite solar loan programs on behalf of our customers in the Two Thirds World.

I listened to the Swiss ambassador give his introductory speech. He noted that two billion people, or 70 percent of the developing world, remained without access to electricity and that they continued to rely on a 19th-century light source, kerosene, which was dangerous, polluting, and unhealthy. He said that solar electricity had proved able to satisfy their demand for household electric power and that for millions of people around the world, solar photovoltaics offered the only solution to their world of darkness. Or some such. I'd heard it all before. In fact, I'd written it! The speech was lifted almost directly from a SELF brochure I'd written a decade earlier. So it goes.

The representative from the Netherlands' Triodos Bank spoke next. Hans Schut and I were partners in SELCO-Sri Lanka and had served together on its board before SELCO bought out the bank's shares, a transaction that produced a small, well-deserved profit for Triodos. They earned it, since they were early risk takers. Now they had joined forces as part of the SDG, which would be co-managed from Zeist, Holland, and Arlington, Virginia. Hans demonstrated for the audience how a "solar light" worked, holding up a small PV module and a luminaire. He then thanked all the international organizations and foundations that had jointly funded SDG, and he recognized numerous officials who came to the podium to offer their own effusive praise for the promise of solar power as a development tool.

Neither Richard nor I were asked to say anything. As one solar investment fund manager had told me earlier from his 27th-floor office on Manhattan's Avenue of the Americas, "You guys are the worker bees. We'll deliver the money." In the international development business, being a policy or grant maker carries a much higher status than being an implementer — one who actually delivers the goods. I thought again of what Indira Gandhi had said.

I liked being a "worker bee." So did Richard and Dr. Hande in India and all the rest of SELCO's 300-plus employees. We generally didn't like this circus train of problematical solar performers who, with few exceptions, were parasites on our industry. But if they had money, we had to learn to love them.

SDG, SDC, SDF — I never knew exactly what entity I was dealing with — sent consultants and staffers to Vietnam, Sri Lanka, and India to look at our operations and analyze our financial needs. These were due diligence trips and necessary, even if expensive and, for us, intrusive and time-consuming. Every foreign visitor had to be taken around in an air-conditioned car or jeep and dined, if not wined, while they probed into our proprietary operations.

The SDG turned down nearly every project and every finance proposal we brought to it, arguing the risk was too great or the "return on investment" too slim or the business model wasn't "sustainable." Our equity investors had taken huge risks and were generally pleased with our progress, and our workers and managers risked life, limb, career, and a premature head of grey hair trying to make this difficult business successful. Yet SDG's unimaginative and immoveable policy and funding bureaucrats and their sycophantic capital management contractors, charged with using public and charitable funds from development organizations and foundations to finance solar business in the Two Thirds World, were afraid to take risks! Their jobs were on the line, of course, and the best way for a bureaucrat to avoid problems is to never, never, never take a risk. In fact, if they avoided making a decision at all, if possible, or deferred all decisions for as long as possible, they would stay employed and comfortable, with an unlimited travel expense account at their disposal.

SDG finally did make us a small loan at market interest rates, which we used as collateral to borrow World Bank funds from the Development Finance Corporation of Ceylon (DFCC) at market rates in order to lend it to our customers in Sri Lanka. As the idiocy of all this became apparent, I grew more cantankerous than ever as we found ourselves unsuspectingly walking the high wire of international development finance at the cirque du soleil.

In 2004, after all the fanfare and folderol, the Solar Development Group folded and was liquidated, and Brooks Brown's EEAF was dissolved. Frustrated staffers had quit as incompetent management replaced original managers who came under attack from funders for failing to make a single loan or investment in the group's first three years of operation. Their excuse was they were busy "analyzing" potential investments, "investigating" possible recipients, and "processing applications" from solar entrepreneurs around the world, including, of course, SELCO-India, SELCO-Sri Lanka, and SELCO-Vietnam. Triodos Bank took over what remained of the portfolio, including our small loan (and, to its credit, made more funds available as it became clear SELCO was one of the few real games in town). Most of the $30 million or so went back to the donors. Solar energy, despite the huge publicity the donors had received for their ostensible support of a sustainable global solarelectric service industry, had apparently proved too risky. Investors were willing to risk their own money, but bureaucrats were afraid to risk *other* people's money.

However, the cirque du soleil was still in business. The Photovoltaic Market Transformation Initiative (PVMTI, mentioned in Chapter 6), launched at Pocantico in 1995, had by 2002 not yet made a long-promised loan to SELCO (we were "short-listed" in 1998). This meant nearly five years of salaries were paid, and first-class airfares and accommodations covered, as the capital managers and project consultants made their way from Washington and London to Bangalore and New Delhi on a regular basis. My comments that the United States entered, fought, and helped win World War II in less time than PVMTI took to make its first loan were not appreciated. I was invited to Marakesh, Morocco, by Mohammed El Ashry, first CEO of the Global Environment Facility, to speak about SELCO at a seminar devoted to the GEF's worldwide renewable energy projects. I chose not go to, in protest.

To understand the cirque du soleil, one must understand the GEF, the $5-billion fund set up by the World Bank and the United Nations, which disbursed funds targeted for environmental projects — including solar and renewable energy — through the UNDP, the IFC, the World Bank, and UNEP. Here were two levels of bureaucracy that employed literally thousands of people in the West; a third level was made up of the consulting firms and project contractors hired with GEF funds through one of the above organizations to "prepare" projects. Project implementation, and the disbursal of actual project funds once all the levels of bureaucracy had published their approvals, generated yet another level of expensive management.

Ironically, these levels of management were staffed mostly by citizens of the Two Thirds World who had made their home in Washington and New York, and who now drew down six-figure salaries. Many had become American citizens. Our apartment building in Chevy Chase was a veritable Tower of Babel, where hundreds of World Bank employees from India, Sri Lanka, Pakistan, Kenya, Egypt, Brazil, Turkey, Nigeria, and Indonesia lived out their comfortable lives, enjoying their careers as "development economists," with no intention of ever returning to what many saw as the godforsaken countries of their birth. Thus, it was likely to be a smug, risk-averse, highly paid Indian bureaucrat in Washington who oversaw projects SELCO sought to develop with the World Bank or IFC.

Dealing with PVMTI, one of the GEF-funded programs run by IFC, constituted the worst experience in our first five years of SELCO operations. We were astonished when Shell Solar India walked away with nearly $3 million in heavily subsidized loan funds because it had the collateral to back it up, while we waited and waited, despite our "priority fast track" status. "They are a blue chip company," the slick young MBA on PVMTI's London-based management team told me, "and SELCO is not." He dismissed the PVMTI founding charter, which had specifically allocated the funds for small entrepreneurial solar companies like SELCO, not the world's second-largest oil company. Mike Eckhart, a longtime proponent of what he called "the Solar Bank," and a well-connected Washingtonian who had formed the American Council for Renewable Energy, called me up, as outraged at this news as I was, to say, "The IFC has just funded the world's largest energy company to defeat the one successful entrepreneur in the rural solar energy business." I hoped he was wrong about the "defeat" part, but I understood the IFC just wanted to make sure it was going to get its money back. We still owed it money for Vietnam.

As I mentioned in Chapter 6, SELCO-India was eventually able to borrow $1 million at extremely favorable terms from PVMTI, in Indian rupees instead of dollars, but nearly five years of wrangling with a brainless bureaucracy had cost us at least $100,000 in real money, not to mention years of lost opportunities. Tens of thousands of poor Indians could have had their SHS financed by PVMTI had the program we proposed been accepted in 1998 instead of 2003. We referred to PVMTI as the Photovoltaic Market Termination Initiative. It takes talented McKinsey-trained specialists to kill markets.

We had placed high hopes on another fund cobbled together by numerous development finance careerists over a three-year period of endless meetings and consultations with the participating agencies. Out of the bowels of the World Bank, the IFC, and the GEF emerged the Renewable Energy and Energy Efficiency Fund (REEEF), which was to invest up to $80 million in renewable energy businesses that would address the developing world's energy shortfall. SELCO was again a "priority" company on its short list. Our colleagues at E&Co and EEAF competed to manage the funds, but in the end, high-level decision makers at the IFC awarded the contract to Energy Investors' Fund of Boston (EIF), a multibillion-dollar investment group with absolutely no experience in renewable or solar energy, and with little interest in it at all — and even less interest in developing countries. But they *were* interested in the accompanying management fees. Millions were spent preparing REEEF; consultants traveled the world and also visited SELCO, which was praised to the skies to the fund's senior management.

The fund never saw the light of day. "Renewables," it seemed, were too risky for a conventional energy investment firm like EIF. The money promised by the IFC, other agencies, and private investors was never disbursed to EIF, other than fees, and not one single investment was ever made. REEEF took up hundreds of hours of SELCO's time as we produced mountains of documentation about our operations, met with its consultants and staff over a three-year period, and squired around its due diligence teams in the field.

Having thrown the first stone at the international development and finance bozos of the solar circus, who seemed to make it their life mission to ensure not one dollar of funds allocated for solar energy was ever disbursed, I must confess that we ourselves were not without sin.

Our biggest failure was Vietnam, where cooperation with the Small and Medium Enterprise program at the IFC resulted in the substantial loan we took to underwrite consumer credit for SHSs in that country. SELCO-Vietnam eventually drew down a large chunk of the $750,000 allocated and was able to pay back a sizeable portion of that. We used much of the loan to finance our customers when the agricultural bank (VBARD) failed to make the loans directly. We were in the solar sales and service business and should not have been a finance company too; then we became a collection agency in order to meet the company's obligations to the IFC, which graciously refinanced and rolled our loan over more than once. Vietnam constituted a huge and expensive learning curve for SELCO as well as for the IFC.

We were also not without sin in Sri Lanka, where we continued to throw stones at the clowns of the cirque du soleil, if I may mix a metaphor. Here, the entire market for SHSs was in the hands of the World Bank's Energy Services Delivery Project, as highlighted in Chapter 7, and it seemed every professional clown in the world had his or her hand in, making our business and our lives as difficult as possible.

Susantha Pinto, with his huge payroll and expanding network of solar service centers, was signing up people faster than he could serve them, faster than Sarvodaya and World Bank loan program could process them, and cash flow continued to be the paramount issue for the company. You have to be paid for the goods sold to be able to order more goods. SELCO-Sri Lanka's sin was to leverage itself beyond its ability to meet obligations to vendors, while focusing on the ever-expanding market, booking sales and reporting income in advance of collecting the funds, and digging an ever deeper hole while sales continued to grow. In business terms it's called "selling yourself to death." Then the Sri Lanka market for SHSs softened as the government embarked on aggressive long-promised, if wholly unsustainable, grid extension. Shell Solar suffered too, as did the many new competitors that entered the market. Sri Lanka continued to be a crucible for the rural renewable energy business, and yet its small market of fewer than a million families was of negligible commercial interest compared to that of the giant to the north, India, with at least 100 million families who needed reliable electricity that only solar power could provide. Nonetheless, Sri Lanka, thanks to the efforts of all the performers in the solar circus, from clowns to fire eaters to lion tamers, may become the world's first solar-powered island, and I will always be proud of helping to produce this circus show.

Meanwhile, in the nonprofit world, the cirque du soleil was an expanding franchise. Donor governments had gotten into solar electrification in a big way, financing huge solar-electrification projects in China, the Philippines, Bolivia, South Africa, and Indonesia. In many cases it was the strong arm of oil company subsidiaries like BP Solar, Shell Solar, and Total Energie that persuaded national aid agencies in Britain, the Netherlands, and France to support rural PV — because it benefited them! No surprise. And although these heavily subsidized projects were largely unsustainable, they did provide solar power and light to hundreds of thousands of needy and deserving families who probably could never have afforded a SELCO SHS, even had we been onsite to sell and finance them commercially. And SELCO was not —

A Chinese peasant cleans his new 20 W solar module in Tongwei County, Gansu Province, in 1993.

yet! — operating in all these many countries where solar projects were proliferating.

The largest was in China, where in 2003 the Dutch government made the Netherlands company Shell Solar very happy by buying some 78,000 SHSs to be delivered, by Shell, nearly for free to peasant and herder households in Xingiang. Neither SELF nor SELCO could compete with that.

Although China had embarked on an ambitious program to bring the electric grid to thousands of remote villages, there were thousands more that it could not reach. I had trudged through many of them in the 1990s when SELF installed over 1,000 SHSs in Gansu Province by 1998, paid for by local commune funds and customer fees. SELF's Gansu project, which began with MaGiaCha in Tongwei County, was funded by the Rockefeller Foundation and, later, the National Renewable Energy Laboratory (NREL) in Golden, Colorado. It was the catalyst for all that came after and the incubator of dozens of new solar companies, government programs, and international donor aid projects in China. Like Shell's.

The heart of SELF's China project was the Gansu PV Company, the Sino-American joint venture of Professor Wang

Anhua's Gansu Natural Energy Research Institute and SELF (see Chapter 2). GPV had its biggest success selling solar systems to Tibetan nomads, who grazed their yaks on the vast grasslands in southern Gansu Province. There are actually more Tibetans in China (in Qinghai and Gansu) than in Tibet, and they are left alone to pursue their nomadic lifestyle. Even their famous and remote monasteries were flourishing, and Tibetan monks at Labareng

Robert Freling

Monastery in Hui Ning County eagerly sought to become distributors for GPV. They sent young monks to Lanzhou for PV training at GPV's headquarters. Soon GPV was delivering solar lighting systems to rural western China by yak-drawn carts.

A Tibetan herder family in Gansu Province is all smiles with the solar electric light and radio in their yurt.

One happy herdsman was featured in a CNN report, standing outside his solar-powered yurt. SELF's Bob Freling, who replaced me as executive director in 1997, fought his way through the foreign affairs bureaucracy in Beijing and Lanzhou to get permission for CNN's Rebecca McKinnon and her crew to travel to the Tibetan areas of Gansu. The report was seen around the world on CNN.

Professor Wang and his son, Wang Yu, focused on marketing their small, reliable SHSs in Gansu through weekly radio "infomercials" broadcast province-wide. But lack of credit continued to plague them, and as GPV discovered, Chinese people, at least those in rural areas, don't do credit. Neither do banks, which in China's Wild West were mostly insolvent and corrupt. This is why the Tibetan market was so important: nomads had money from selling livestock, and they paid cash. They also knew no government program would ever connect them to the grid since they moved around, as nomadic people tend to do. They were perfect candidates for portable solar PV! Our Chinese farmer market, however, became increasingly difficult to service. Peasants remain among the poorest people in China because the government sets agricultural product prices to benefit cities, not the farm economy, and because corrupt local officials heavily tax and otherwise oppress farming communes every way they can. Without credit, or heavy subsidies, farmers could not afford solar. SELF could subsidize SHSs, and so could the government, but SELCO could not, which is why the company never made a serious play to take control of GPV, even though we recognized China was the world's second-largest market for PV after India.

The half-million-dollar NREL project saved the Gansu PV Company, at least for several years, by providing subsidies to farmers purchasing PV systems. I had instigated it by inviting NREL's PV specialist William Wallace to visit GPV and our "pilot villages" in China, and even though he was overweight and out of shape, he managed to trek into some pretty remote places and up steep mountain paths with Professor Wang to see how solar electricity was transforming lives in rural China. After long negotiation with all the parties concerned, Bob Freling worked with Bill Wallace to nail down SELF's biggest government contract to date: a cost-shared project between the US Department of Energy and the Ministry of Science and Technology in Beijing.

As a result of our efforts and our countless trips to China and endless meetings in Beijing, Lanzhou, New York, and Washington, we had inadvertently launched a Chinese cirque du soleil. Everyone wanted to join up. Bill Wallace later headed the UNDP's renewable energy office in Beijing and began talking to the Chinese government about "institutionalizing" solar photovoltaics and integrating PV into its rural and agricultural programs. Soon I found myself speaking at numerous PV conferences in China, and Chinese officials began making their way out to Gansu to see MaGiaCha in Tongwei County and to visit Professor Wang at GPV.

GPV was now famous. A Beijing taxi driver told me he had even seen a story about it on national TV. Back in Washington, I would bump into colleagues who had just returned from China to study solar, and they too had ventured out west to industrial Lanzhou to visit China's famous pioneer of solar electricity, Professor Wang. American solar experts Debra Lew and Peter Lowenthal went, and Dr. Chris Sherring, an NREL subcontractor who published fat "China PV Business and Applications Evaluation" reports that featured GPV and monitored a growing industry comprising some 400 small solar companies in the world's most populous nation. Even high-level government officials, like Mrs. Deng Keyun, former head of rural energy for the Ministry of Agriculture, were now "associates" of American NGOs like Winrock International, which opened a Renewable Energy Project Support Office (called a REPSO in Washington's energy NGO culture). Winrock, with an enormous budget from USAID, sent legions of "experts" to China, which clearly did not need anyone to tell it how to promote renewable energy, and recruited more Chinese to work for Winrock instead of the government or private firms, rewarding them with trips to the United States to attend more solar conferences. The cirque du soleil boasted three rings by the end of the millennium, with the center ring devoted to China.

I sent my own consultants to assess the business prospects for GPV, which by 1999 had reportedly installed some 7,000 small SHSs. I wanted to see if it was worth bringing GPV into the SELCO fold, since we had acquired SELF's 49 percent ownership of the company. Some of the people who I sent to join the circus were: Charlie Benoit, now living in Shanghai, whom I'd known in Vietnam during the war; our own Jon Naar, formerly of USAID (see Chapter 6); Daniele Guidi from Italy, whose Tufts University's graduate thesis on "Solar PV for Developing Countries" launched him on a lifelong career in solar electricity; and Scott Vaupan, a China specialist who had done exhaustive studies on renewable energy markets in China for NREL and the World Bank.

However, in 1999, despite the booming economy in Gansu, the stability of the yuan, the right of foreigners to own more than 51 percent of critical industries like energy, the constitutional recognition of private enterprise, the newly official status of capitalists in China, and all the glowing reports about GPV, I remained skeptical about the wisdom of investing SELCO's capital in GPV. I wasn't getting clear information from my own visits or from Bob's or Charlie's, even though both of them spoke fluent Mandarin (I only knew "hello" and "thank you"). I later learned that in those days — maybe it is still the case — many Chinese were actually offended when foreigners dared to learn their language, making it impossible for them to talk behind our backs, make fun of the "long noses," and otherwise scheme against our interests. It was better, in many cases, to work with an independent interpreter, and even then, Professor Wang often became incensed when I brought my own instead of using his, since he could not "correct" the translator who might actually tell me the truth. Fortunately, Professor Wang spoke just enough English that he and I could communicate directly, and his communication skills improved considerably during rounds of maotais at the nightly banquets. (He learned his halting English while under house arrest during the Cultural Revolution of the 1960s.)

My concern was that GPV was not growing at the rate the market warranted, despite the credit issues. Quarterly financial reports, with official stamps from the provincial tax authorities, always showed losses. Professor Wang made lots of sales, but revenues never exceeded costs, although I knew his 36 employees were not costing him that much, and he owned his workshop and offices. Imitators had sprung up on Professor Wang's doorstep — which he tried hard to hide from me. Many were selling technically inferior products. Several former employees, as well as people he had trained earlier at the UNDP-funded Gansu Natural Energy Research Institute, started their own small companies, and because they were younger, less academic, and more businesslike, they soon had large PV concerns of their own, selling to the huge cash market of nomadic herders in Qinghai Province next door to Gansu.

As Professor Wang explained to me one day, "The largest social institution that works in China is the family." He trusted no one outside his immediate family, and he was always cutting strange investment deals with relatives, illegally importing Toyota pickups for his company from Hong Kong through cronies and family associates, and generally being circumspect about his banking and government relationships, which grew more confusing by the year. At the same time, whenever he brought in an outsider to handle rural product distribution, he was invariably ripped off, another fact that only Bob and Charlie could ferret out. This same phenomenon exists in Vietnam, where there is nothing between family businesses and state-owned enterprises, or between family companies and foreign offices of multinational corporations. Small to medium-sized independent corporations simply do not exist in China and Vietnam, and nothing in these societies is transparent.

Nonetheless, GPV trucked along from its fifth-floor offices in central Lanzhou, which SELF helped buy for about $80,000 in 1995, carefully preserving the bulk of the original capital provided by the Rockefeller Foundation through SELF. GPV perfected a "cash-and-carry, plug-and-play" portable SHS that required no service network to install or maintain. By 2000, most sales came from peasants who heard GPV's radio ads, took a bus from their remote communes in rural counties, found their way to the office in the huge, dirty, provincial capital on the banks of the Yellow River, and walked up the five flights to make a cash purchase of an SHS. The product was explained to them — how it worked and how to install it. Then it was

A Chinese peasant with his "cash-and-carry, plug-and-play" SHS from Gansu PV Co., co-founded by SELF with funding from the Rockefeller Foundation.

boxed up, shrinkwrapped, and hoisted onto their backs, and off they went as solar pioneers in their own right, the first in their town with electric light.

GPV's customer lists became its marketing program, and every customer was recruited to bring in more buyers; prizes and incentives were offered, mirroring the Amway sales approach. Peasants became GPV brokers. Testimonials were the preferred form of advertising, used on the radio and even in several television specials sponsored by GPV. For marketing "collateral," GPV printed tens of thousands of pamphlets, flyers, and slick color brochures describing its products. Another element of GPV's marketing program was to hand out promotional cigarette lighters with the Anhua SHS logo because without the always-lit kerosene bottle lamp on the family table, the men had no way to light their cigarettes. This was the only downside of owning an Anhua SHS.

Wang Anhua wanted to appoint a broker in each village, following Mao's "people's war" precepts, and called this his "thousand points of light" marketing strategy. Besides the walk-ins, he soon had letters from 2,000 people who wanted to purchase an "Anhua SHS" (as he called his small plug-and-play systems), but who were not sure they could afford one.

"Mr. Wang, I am a farmer living in a remote village on Wushan Mountain. Every time I see people in the valley below with lights in the villages, every time I see them watching TV while they have their supper, I become very envious. Although we have enough to eat in this village, our sources of income are still rather limited. At present we have no way of coming up with enough cash to bring a power line to our village. Therefore, I am thinking to use my own money to purchase one of your SHS. Respectfully, Wang Shucun."

From Liu Chunming, Neiguan town, Lin Chuan village, Dingxi County: "Hello! In the small town where I live, we often suffer from electric blackouts. For example, we completely missed the Hong Kong Handover on July 1 [1997] due to a blackout. Several days ago I heard a radio advertisement discussing a solar device which can 'borrow the sun to light up the night.'" The writer goes on to ask numerous questions, then says, "If the price is reasonable, and it is easy to use, my family wants to buy one. We can also promote this technology to our neighbors."

Zhou Xiangyi wrote from Ming County: "I am a farmer living in a mountain village. I was recently appointed as head of my village. The other day I heard a radio announcement discussing how your 'Anhua' solar home system can provide light to unelectrified villages. This news made everyone in our village very excited. We would like to purchase 30 SHS." The writer then proposes they make a down payment and then installment payments "until the units have been completely paid for. This is a poor village. We do not have much money."

And from Nan Lianxi in Qingshui County: "After hearing about your solar home systems from the radio advertisement, I became very interested. All 50 homes in our village are without electricity. At night we have to live under the dim light of kerosene lamps .... Due to the fact that solar electricity involves a highly advanced technology, it has not been widely used so far. Many farmers in our village have not even had a chance to hear about solar electricity, so it is difficult for them to become consumers." After asking a lot of questions he writes, "We are now in the process of harvesting our wheat, so everyone is very busy. After the harvest is complete, I would like to travel to Lanzhou to discuss everything face to face."

"I wanted to watch TV before I died," Luo Yanzhen told a reporter from the *Far Eastern Economic Review*. A GPV customer from Niucha village, 180 kilometers southeast of Lanzhou, she also bought a black-and-white TV to go with her small PV system. "I've seen Premier Li Peng and President Zhang Zumin for the first time." The weathered farmer had sold ten small sheep and eight baby pigs to purchase a solar home system, "but it was worth it," she said.

Over the years, I personally listened to hundreds of personal testimonials from peasants as they described how electricity changed their lives, how they no longer needed to use kerosene, how much better their children could study at night, and how much they enjoyed seeing television for the first time. One I recall was an old man who said, having seen Zhang Zemin on TV, "I have never before been able to see the emperor!" Another elderly peasant, shedding tears, said, "I have long heard that city folks do not need oil to generate light, but in all my 70 years, this is the first time to actually see such a phenomenon with my own eyes. What a beautiful thing!"

Wang Anhua encouraged these written testimonials by offering prizes and free trips to Lanzhou, and one customer wrote, after installing an SHS, "As the afternoon wore on, the house filled with more and more neighbors from the village who continuously turned on and off the SHS and asked all sorts of questions. Expressions of envy as well as expectation were clearly visible on their faces. At that moment, their hearts were illuminated with the light of hope. The next day another family in Xiping village decided to purchase a solar home system. They have a son who is about to take an examination to enter high school. The parents hope that the bright light of the SHS will help their son better prepare for the exam. The parents have invested a great deal of hope in their son, and now they are also putting hope in solar electricity. Suddenly, I feel as though this year's sun is especially bright, and this year's winter especially warm."

But the best came from Liu Xueming in Huining County: "On the afternoon of October 4, I returned to my village, carrying a television set, radio, and a large cardboard box. I thought something was wrong with me because everyone was staring at the mysterious box I was carrying. It was 4 PM. I removed the solar panel from the box and connected the positive and negative wires to it as indicated in the instructions. Immediately, a green LED light lit up as the wires were connected. Gasps of amazement came from the room. How could a glass panel turn on a green light, just like that? I laughed and explained the green LED indicated that the battery was being charged by the solar panel. I told everyone to be patient and that in a few minutes they could watch television. As I turned to connect the TV set, I could hear a few older people murmuring among themselves. 'Don't be so boastful! How could a small glass panel, exposed to a bit of sunlight, allow us to watch television? Don't pull the wool over our eyes just because we're a bunch of old country bumpkins!'

"I proceeded to turn on the TV set. That's when everyone became very excited. They hugged one another, and shouted with glee, 'We have electricity in the village! We can watch television!' My own parents asked everyone to take a seat and get comfortable. The older ones sat on the sofa, the children sat on the kang. Everyone sat there watching TV with huge smiles on their faces. My wife poured tea for the older ones. We watched a soap opera, and after it was over I turned on the electric lights, and again exclamations of surprise filled the room. 'How much brighter than our kerosene lamps!'

"I am 25 years old, and until this day I had always lived with kerosene. My son was born a few weeks after I brought home our solar home system. Our home shone so brightly that night! As relatives and neighbors came by to congratulate us, I could not help but reflect on my good fortune. Everyone said to me, 'Xueming, you really pulled it off. You went to Lanzhou and came back with lights and a television! This is something we have never seen in all our years.'

"In honor of our good fortune, I decided to name my child 'Guang Dian'." Guang Dian means photovoltaics in Chinese.

My friend Anil Cabraal from the World Bank visited Wang Anhua, GPV, and our pioneer solar villages, the first in China, in 1996. I began meeting with World Bank staff in Beijing during my visits to China, and with Anil and his colleagues at the Asia Alternative Energy Unit back in Washington. Streams of World Bank, UNDP, and GEF consultants made their pilgrimage to GPV to see how this little company was commercially able to manufacture and sell such small solarelectric systems which, thanks to Wang Anhua's electrical engineering skills, were not only reliable, but also amazingly efficient. (I won't get technical here, but this means they produced a lot of light, and even spare power for a small black-and-white TV, with a mere 20 Wp Chinese-made solar module. Wang used solar modules from Ningbo Solar and imported sealed gel-cell batteries from Panasonic in Japan.)

"I don't see how he gets his fluorescent lights to work on such low amperage," Anil told me. I didn't either. Anil was amazed to see GPV's small, attractive, red wooden box that contained the battery, charge controller, switches, and easy-to-read meters and was entirely transportable and very rugged.

In 1997, GPV was listed among 17 solar companies qualified to participate in the "upcoming" World Bank solar energy program; it was the only Sino-American joint venture — the rest were entirely Chinese. We had high hopes this program would provide the subsidies poor peasants needed to acquire solar electricity. It would be channeled through commercial entities, as had been done in Sri Lanka.

Five years after the program was announced, it was launched in 2001 — after Professor Wang had fallen into ill health, after his son had proven incapable of managing the company, and after SELCO had decided GPV couldn't be rescued with any amount of money. We looked at partnering with some of the aggressive new upstart PV companies run by young Chinese entrepreneurs, but they explained they didn't need any Western capital. And although they all knew and respected Professor Wang as the true pioneer of solar electricity for rural people in China, no one wanted to buy GPV or its technology. Not when they could steal it (no copyright, patent, or other proprietary protections are available in China, where much of "capitalism" might better be described as rampant piracy).

The Darden School of Business Administration at the University of Virginia did a case study on SELF in China in 1997. It reported that SELF had been a catalyst for rural PV electrification in China and that its operating plan was always to generate self-sustaining commercial enterprises and associated credit funds. The report stated, truthfully, "as a market began to form, SELF would exit China, thereby allowing the forces of the market to electrify the unelectrified rural areas." Now it was happening, but without SELCO.

In 2002 the World Bank put $205 million into the China Renewable Energy Development Project, a quarter of it earmarked for PV, to partially fund some 350,000 SHSs. The first 100,000 have already been installed and have qualified for a $1.50 per watt World Bank subsidy that can be captured by the vendors. Frankly, having learned in 1999 of the decision to provide a subsidy that would work out to only about $30 per $300 system, I wasn't sure the pain and anguish of dealing with the World Bank's in-country bureaucrats would have been worth it, but we'll never know. Other companies are now dealing with the cirque du soleil in Beijing, which is also overseeing quality control, technical standards, market development, product design, cost-shared grants, and whatever else these well-intentioned solar circus performers can get their hands on, while imposing an "accreditation" process that will drive entrepreneurs nuts.

Eclipsing the World Bank project is the Chinese government's own ambitious Chinese Brightness Program, which aims to deliver solar power to one million peasants in Tibet, Yunan, and Sichuan. This was jointly funded by China's ministry of finance and Kreditanstalt fuer Wiederaufbau (KfW), the German state-owned development bank. In 2001 the State Planning Commission instituted its "renewable portfolio standard," requiring a substantial percentage of provincial power generation to come from renewables, including solar power. It also made funds available to China's many solar manufacturing facilities, mostly owned by one or another government ministry. And in 2004 the State Development and Reform Commission announced its own program of "solar energy exploitation" to the tune of $1.4 billion!

In mid-2004 the *China Daily* reported that severe environmental problems resulting from the country's reliance on coal, coupled with serious energy shortages, were driving the National People's Congress to consider massive support for renewable energy. (In addition to pollution, coal was dangerous in another way: 6,000 miners died in Chinese coal mines in 2004.) In June the Chinese government announced at the World Renewable Energy Conference in Bonn, Germany, that it would meet 10 percent of its energy needs from renewable resources by 2020. "Priority will be given to developing renewable energy in rural and remote areas to meet the lifestyle and work demands of local people," the government stated. SELF had arrived in China ten years too early.

By late 2004 China was experiencing near-catastrophic energy shortages and expecting a power shortfall of 30 million kW. At the same time, the country's skies were perpetually hazy from burning carbon fuels. Some of these particulates have been identified falling on California. By the end of this decade, China could be the world's largest producer of greenhouse gases and carbon monoxide.

From polluted Beijing, *New York Times* columnist Thomas Friedman wrote in June 2004, "Developmentally, China's growth is soon going to be restrained, if it isn't already, by a sheer shortage of energy. Strategically, China and America could soon find themselves in a dangerous head-to-head competition for fuel." He called for "big imagination," for the US president to propose "a grand China-U.S. Manhattan Project — a crash program to jointly develop clean alternative energies, bringing together China's best scientists and its ability to force pilot projects, with America's best brains, technology and money." He then quoted Scott Roberts of Cambridge Energy Research Associates: "When it comes to renewable technology and sustainable energy, China could be the laboratory of the world — not just the workshop of the world."

In "The Chinese Century," a July 2004 article in *The New York Times Magazine*, Ted Fishman claims that "China is getting ready to supplant the U.S. as the capitalist engine of the world." Napoleon observed, 200 years ago, "Let the Chinese giant sleep, for when it awakens, the world will tremble." We isolated Mao and his communist revolution, but we can't isolate Chinese capitalism. China will soon become the largest user of energy in the world, and what they do there will affect us all.

I'm not writing a policy book, prescribing what "must" be done if the world's environment is to survive the American century, let alone the new Chinese century, but it's clear — and becoming clearer to more and more Chinese policy makers — that the Chinese century had better be a solar century.

At the dawn of what we all hope will be the "solar century," SELF, now long out of China and under Bob Freling's leadership, concentrated on South Africa, Brazil, Nigeria, and Bhutan.

Picking up on the Valley of a Thousand Hills solar-electrification project I'd begun many years before in South Africa, Bob helped the Zulu community of Mapaphethe close the "digital divide" with a pilot solar project centered on Myeka High School. Solar power supplied by SELF allowed the school to operate a dozen computers contributed by a partner NGO, while SELF provided a satellite uplink and receiving station powered by the sun. Students could now access the Internet in a community that, until then, had only one solar-powered telephone. Soon they had their own website and were "plugged in" to the global solar-powered schools program, which looked to Myeka as a beacon of light and hope. Two students from Myeka High School later attended a global "bridging the digital divide" conference in Mexico City.

SELCO planned to launch a subsidiary in South Africa, and we registered the name and formed a partnership with a local entrepreneur. The government, however, simultaneously announced plans for a 90 percent subsidy for

SHSs. This progressive-sounding renewable energy program unintentionally killed the private solar market in South Africa. SELF, ironically, had been instrumental in getting the Mandela government to focus on SHS as a complement to the aggressive electrification program of Eskom, the country's power utility. Two million homes that would not be connected to the grid were officially targeted for solar — and then nothing happened for five years, aside from a failed program instituted by Shell and Eskom, with which SELCO could not compete. Only now has solar electrification taken hold in South Africa with investment from large international utility companies.

SELF electrified a coastal community in Ceara, Brazil, north of Fortaleza, in the 1990s, and in 2002 it went back, this time to the Amazon, to bring electricity to a village located in the Xixuaú-Xiparina Ecological Reserve, 40 miles up the Amazon from Manaus. Here, SELF installed a solar power plant large enough to operate an OnSat satellite dish, computers, lights, and a vaccine refrigerator. Now the Caboclo people of this remote Brazilian rainforest preserve can access information through the Internet and communicate with the outside world for the first time. Always the promoter, Bob brought along a video crew to chronicle the project, producing an excellent documentary for television.

Bob then took SELF into one of the toughest nations to deal with in the world — Nigeria. Africa's most populous country (130 million people) has three things besides lots of people: oil, corruption, and no electricity outside the cities. And it has sunshine. The governor of Jigawa, a state in the dry, Muslim north near Kano, asked SELF to bring solar power to three villages that had never seen electric light or electricity. USAID and the US DOE stepped in, matching funds from the state government, and SELF launched a half-million-dollar pilot solar project, the largest of its kind ever attempted in Nigeria. SELF hired Jeff Lahl, an experienced electrical engineer and project manager, to reside onsite until it was completed. Governor Ibrahim Turaki proudly inaugurated the project in early 2004, with a CNN news crew capturing on video the delight of villagers who now had a source of power to run a water purification system, mobile irrigation pumps, streetlights, lighting and computers in their schools, a vaccine refrigerator for their health clinic, and lights for the surgery and three mosques. As well, a 1,600-watt PV array powered a unique micro-enterprise center in a one-story cinder-block building where six small businesses leased space. Sixty householders also signed up to purchase SHSs. All this was powered not by oil, but by the blazing, relentless, African sun.

CNN ran the story for a month worldwide. President Obansanjo visited the project. Three other Nigerian states asked SELF to help them do similar projects. One hopes that Nigeria will plow its oil profits into bringing solar power to its people, who have gained little from selling off the nation's wealth. Why should foreign donors have to do this when Nigeria is so rich? Where is Shell Solar, given that Shell is the largest exploiter of Nigeria's oil resources?

In Bhutan, SELF found a partner in the Royal Society for the Protection of Nature, which helped subsidize the costs of 151 four- and six-light SHSs close to a nature preserve for endangered black-necked cranes because installing overhead electric lines would be hazardous to their health. Not having electricity was also hazardous to the health of Phobjikha's residents, as they had relied on kerosene lighting for the past century, replaced by wood when they could not get kerosene. SELF installed a 750-watt PV system at the nature center's headquarters to run a computer and other office equipment.

In a rare cooperative venture between the two organizations I had started, which had long since gone their own ways, SELCO-India was contracted to supply all the systems, do the installations, and send technicians to train seven local youths in solar system maintenance. The equipment and materials were shipped by truck from Bangalore all the way to Bhutan. Following SELF's modus operandi of not giving away systems, as so many failed donor programs had done, and instilling a sense of responsibility (SELF-help), the villagers in this remote hamlet were asked to pay part of the cost of their SHSs on a three-year installment plan. "I hope I can pay for the installments from the sale of potatoes," said Tshewang Zam when she turned on her lights for the first time. "Besides that, we can also stitch our embroideries for sale, we can do weaving at night, and our children can study under better conditions."

Bob attracted the rich and famous to SELF's board, including actor Larry Hagman, Templeton Prize-winning physicist Freeman Dyson, and a wealthy San Francisco couple, the Swigs. Ed Begley Jr., the actor at whose Studio City house I had held a SELF fundraiser in the early days, joined the board. Ed was everybody's environmentalist; he talked the talk, walked the walk, lived in a solar-powered house, cooked vegetarian meals on a solar cooker outdoors, and drove an electric car. With a few more Ed Begleys around, the world would be a better place. Larry Hagman held fundraisers at his mountaintop aerie in Ojai, California, and Freeman Dyson opined for the national media on how SELF was changing the world. I wished. But if Freeman Dyson believed it was, I wouldn't object.

The charitable efforts to light up people's lives with solar electricity continued apace around the world, but what were often termed "do-gooder" projects were being eclipsed by commercial approaches to solar rural electrification as more and more entrepreneurs got into the business. Nonetheless, even giant Shell and its subsidiary, Shell Solar, had to find subsidies for half their customers. New groups like Green Empowerment in Portland, Oregon, built a network of worldwide projects on SELF's model, but expanded into other small-scale renewables to provide a wider array of energy services to rural communities. Solar pioneer Richard Hansen wrote in mid-2004, "Two billion people — a round statistic for a sharply divisive global problem: with several hundred million families worldwide still relying on crude energy sources like kerosene to light homes and dry-cell batteries to power radios — at 500 times the price of grid electricity — the 'energy divide' casts shadows over much of the globe."

The *energy divide* was even more critical than the digital divide, but few recognized it, especially in a world where oil prices were soaring and people living the good life in America and Europe were now worrying about their own energy supplies. Chinese cities were worried about keeping the lights on, and in 2003 the lights went *off* in the northeastern United States because we had failed to use energy efficiently and peak demand overwhelmed the grid. Would the Two Thirds World emerge from darkness before the so-called First World plunged into it?

Greenpeace USA and Greenpeace International launched a global solar campaign in 2001, a decade after I left my media director post at Greenpeace to start SELF. In "Power to Tackle Poverty," a paper outlining a "strategy to deliver renewable energy to the world's poor," a Greenpeace campaigner wrote: "As we begin the 21$^{st}$ century, one-third of the world's population are still living without access to electric lighting, or adequate cooking facilities. Providing these two billion people with modern energy systems that are able to meet their basic needs for clean water, health care, heating and lighting is *one of the most pressing problems facing humanity today*" (italics in original). Noting that the G8 had set up a renewable energy task force "calling for 800 million people in developing countries to be provided with energy from renewables within ten years," Greenpeace pointed

Global Green USA

out that this would still leave "over one billion people dependant on unsustainable energy sources, resulting in severe environmental and social impacts." Interestingly, the chairman of the G8 renewable energy task force (on which I served as an advisor) was the recently retired chairman of Shell, Mark Moody-Stuart. In our occasional

SELF'S Executive Director Robert Freling using his fluent Russian to speak with Mikhail Gorbachev at Green Cross Millennium Awards.

e-mail exchanges I never summoned the courage to ask him why — why him? How did he get from chairing the world's second-largest oil company to chairing an international forum on renewable energy?

In June 2004, former Soviet prime minister Mikhail Gorbachev committed to "fight energy poverty for two billion people without electricity" and called on the Bonn Renewables Conference 2004 to adopt his plan for a $50-billion "Global Solar Fund." Gorbachev's Green Cross organization, which had conferred the Millennium Award for International Leadership on SELF in 1998, pointed out that such a global financial commitment to solar power would lower the cost of photovoltaics, address "energy poverty," deal with urban peak demand (which caused the northeastern US blackout), and create an "energy glasnost" to open electrical grids to flexible, decentralized, smart energy solutions.

If Gorbachev could bring down the Soviet state and end communism almost single-handedly (no, I don't think Ronald Reagan had that much to do with it, but I won't argue this here), maybe he could be effective in helping to close the energy divide. Unfortunately, as I write this, the Global Solar Fund is just another three-ring attraction under the Big Top, looking for an audience at the cirque du soleil.

It had been a long time since I was invited to the top-floor boardroom of the Shell building in London, overlooking the Houses of Parliament and the Thames. Shell executives had asked me to come to London in 1998 on behalf of the Solar Electric Light Company, launched only the year before, so, in their words, "we can make you an offer you can't refuse." Shell Renewables had just been formed, and Shell's manufacturing plant in Helmond, Netherlands, had been renamed Shell Solar. For the first time a huge multinational had put its brand name directly on solar power technology and products, and now it wanted to brand "rural solar" in the developing world. We'd already branded it, with SELCO's bright red sunburst logo, and Shell wanted a shortcut to these hard-to-reach markets that we'd spent so many years cultivating. It wanted to avoid our painful years of trailblazing and get right to some of those two billion without power, now.

"We could call the merged company 'Shellco'," they joked.

Neither I nor SELCO's investors had any problem with "selling out" to a multinational that had the deep pockets to get the job done and making a small profit for our early-stage risk taking. But SELCO had not yet even begun to create value, so at this stage it wasn't worth very much. Which of course is why Shell didn't offer very much. In fact, the offer was so ridiculous that SELCO's board broke up laughing when they heard it. "I thought career executives in multinational corporations were smarter than that," I said. "What's the point in making us an offer we can't refuse when common sense would tell you that we would refuse it out of hand?"

Not long after, as I've mentioned in Chapters 6 and 7, we found Shell in our face in Sri Lanka and India, talking to the same bankers with whom we'd crafted business relationships and tapping communities where we'd worked hard to make inroads, which of course made for productive competition and the creation of a real market for rural solar. And Shell Solar was awarded the aforementioned $3 million in low-cost working capital by PVMTI, which had not yet lent us dollar one.

The Rural Solar Division of Shell Solar came under the leadership of Damian Miller (who was not the executive whose offer we refused). Damian turned out to be a crackerjack executive, overseeing complex and far-flung operations in Asia and Africa from his base in Singapore, and giving SELCO a run for its money. Six years after the short, happy meeting in London, Damian requested my support for the Million Homes initiative he had proposed to the Global Environment Facility in 2002. Although we were competitors, we were also friends, harking back to the time at SELF when I opened our doors around the world to assist him with his Cambridge University PhD on "solar electrification in the developing world." Now we represented the two largest companies in this new field, and he needed SELCO's support for his plan. His idea was that G8 countries would make an additional combined donation of $150 million to the GEF to subsidize SHSs at the flat rate of $150 each. Having witnessed what worked and what didn't work in the solar energy world, he had proposed a fair and straightforward disbursal mechanism that would allow commercial entities to install one million SHSs in five years. He wasn't suggesting the World Bank kick in any money, only that donors to the GEF put up an additional $150 million targeted exclusively for this program — almost exactly what the United States was spending *per day* in Iraq.

With the support of Phil Watts, Shell's chairman at the time, who had personally called World Bank president James Wolfensohn to suggest the idea, Damian invited me to several meetings in Washington in the spring of 2003 to explore this breakthrough solar-electrification proposal with officials at the GEF, the IFC, and the World Bank. Before you could say "three-ring circus," a dozen consultants were hired, "expert working meetings" were convened, analysis was undertaken by World Bank contractors, thick reports were generated, and recommendations were made. And then one of the cirque du soleil's ringmasters took centerstage. In 1999 the World Bank managed to find a bland, unimaginative bureaucrat with no relevant experience to take charge of its Solar Initiative. When his name was announced, I was stunned, for Greenpeace UK had already investigated the candidate and proclaimed him unsuitable for this key position. He would be, like it or not, the UK's "contribution" to the Solar Initiative, as that country would be paying his salary at the Bank. His presence was poison to all projects and proposals he came in contact with. It was as if someone high up at the Bank had decided that solar power had to be killed and hired a hit man, with tenure, to kill it.

It didn't take more than two months for the official in question to kill Damian's and Shell's Million Homes Initiative — even though James Wolfensohn thought it was a good idea — and it remains deader than road kill.

Solar circus performers never say die, and shortly thereafter a new initiative was launched with World Bank pilot funding (meaning salaries, consulting expenses, and airfares, but no program money). It was organized by solar energy's perennial insiders, who to my knowledge had never funded or managed or implemented an actual solar project, but who knew where the money was and how to keep the solar circus show going. The new group was the Global Village Energy Partnership (GVEP), with seemingly limitless funds for airfares and meetings. Since only about 50 weeks of the calendar year were filled with solar energy conferences, GVEP filled up the other two and flew in world experts for yet another global gabfest in Manila or San Francisco or Washington on how to bring renewable energy to the world's poor villages. To make sure they got expert advice, they coopted our own Dr. Harish Hande, put him on the board, and invited him to be a keynote speaker. I remain skeptical, and so does Harish, but maybe more talk will lead to action. After all, our company, and a good portion of the cirque du soleil, was born at Pocantico, a landmark three-day talkathon. But sometimes I can't help but see some of these well-intentioned energy organizations as infected with meetingitis — parasitical at best, outright exploitive at worst. To date, no money for solar or renewables has emerged from the activities of GVEP, but it keeps on "conferencing."

And this brings us back to the World Bank, which funds so many of these useless gatherings and "initiatives," only a few of which I've highlighted. Accountable to no one — for no single government dares to take it on — the World Bank, along with its partner institutions, the International Monetary Fund (IMF) and the International Financial Corporation (IFC), has appointed itself the bringer of economic success and stability to the Two Thirds World by keeping half of Asia, Africa, and Latin America indebted to it. The "Fifty Years is Enough" campaign in 2004 marked the 60th anniversary of the World Bank with ongoing protests against the debt crisis, the structural adjustment fiascos, the lack of transparency at the institutions, and the destructive dam, oil, and other infrastructure programs. The Bank is still "mortgaging the earth."

Even the Bank recognized the problems with its infrastructure programs. Its Extractive Industries Review recommended that $800 million in annual lending to oil and mining interests be phased out because these industries cause environmental destruction and climate change, while exacerbating poverty instead of alleviating it. The bank's management and the oil companies were very unhappy with these findings, which noted that 94 percent of Bank lending went to fossil energy, while only 6 percent was used for renewables, mostly hydro (less than 1 percent for solar). The Bank's own internal review urged that the $800 million be redirected to renewable energy.

Adding to the Bank's woeful reputation was the US Senate Foreign Relations Committee hearings chaired in May 2004 by Republican senator Richard Lugar. Witnesses testified that "$100 billion may have been lost to World Bank corruption." I can only say that I've witnessed the corruption on a very small scale in developing countries, where the elite line their pockets with World Bank project funds and send their children to private schools in Europe and the United States, where, with their newly minted PhDs in economics, they can get a job with ... the World Bank! It's the economics of self-preservation in what might better be called the deteriorating world, planet preservation be damned.

In 1993, when I first approached the Bank to support solar, I was aware even then of a leaked internal World Bank report that claimed one-third of all Bank projects were failures and that the failure rate had increased 150 percent over the previous ten years. In 2004 the Poverty Action Lab at MIT, founded by a group of development economists, evaluated the Bank's progress and determined there was "scant evidence" that its myriad projects have made any real difference in improving the lot of the world's 2.7 billion poor people. So what else is new? Sixty years *is* enough.

The SELCO group, meanwhile, was prospering in the difficult retail markets it had chosen to focus on in India, Sri Lanka, and Vietnam. It was a survivor. Two weeks after we raised our initial capital in the go-go 1990s, the

Asian financial crisis hit — heralding the bursting of future bubbles, like dot.coms, that would go the way of Bangkok real estate — but it had little effect on the company. SELCO survived a global downturn in commodity prices that affected the ability of our customers to sell their crops and products. SELCO-India survived, and is surviving, globalization that flooded India with cheap goods and comestibles from China. SELCO-Sri Lanka survived the ongoing war in that country, although it suffered a blow from the Christmas tsunami of 2004, while SELCO-India survived the Enron debacle of Dhabol and Enron's own demise, which together gave energy markets and American companies a big black eye. SELCO survived the implosion of the stock market in 2001, which left millions of Internet company investors holding worthless stock, and the corporate scandals that left millions of shareholders with nothing. SELCO survived September 11, the Iraq War, and the near collapse of the dollar under George W. Bush, and it is still going strong, I am proud to say.

The only battle we couldn't win was in Vietnam (SELCO is still fighting on that front, however). In Vietnam, no one defeats the communists, not the full weight of the US military over a decade of war, not globalization, not a booming economy, which is anything but communist, and not SELCO. No one "beats" the Vietnamese at war or in business, as SELCO learned the hard way. They are the most stubborn people on Earth, and they will have it their way, whatever that "way" may finally be. Maybe their way is the best way, and who can challenge that? It's their country.

I stepped down as CEO in 2002, for I believed it was time to find a younger CEO and to begin the process of pushing the overall managerial functions down to country operation levels — SELCO's key to success was trusting local foreign nationals to do the job. I finally left the board and resigned as a director in 2004 after new investors bought out our largest shareholders and took control of the company. Their deep pockets suggested that SELCO would not be without financial backing for growth and expansion in the future. Harish Hande, a natural leader, eventually stepped into the top position.

In 2002, SELCO-India made its first audited profit. "I can't believe we're actually making a profit selling solar energy to poor people," Harish Hande told me. The profit was slim, and a long time coming, but it was there.

The Solar Electric Light Company finally reached break-even in 2004. The company continued to focus on its objective of meeting not only Wall Street's rigid bottom line, but the Triple Bottom Line as well: social responsibility, environmental sustainability, and economic profitability.

By the start of 2005, SELCO had sold and installed over 50,000 solar home-lighting systems. No other company had achieved this. It was selling energy services to underserved populations, and its revenues were growing at 35 percent a year. It was neither extracting resources nor exploiting workers, as happens too often with foreign direct investment. SELCO sought to reduce energy poverty among people in need, not exploit energy poverty as Enron had tried to do. We had proved the market economy could serve people first, and by serving people we built a successful company.

SELCO's 350 workers and managers had a never-say-die, can-do attitude. The future belonged to them, if not to the company itself. But since this was capitalism, not philanthropy or socialism, the company belonged to the shareholders, not the workers, and we had to follow capitalism's rules, including the Golden Rule — The Person With The Gold Rules. Most of our shareholders were exceedingly patient; they had invested in SELCO knowing full well that this was a long-term proposition requiring more than usual patience. Both Rolf Gerling and Stephan Schmidheiny were patient (billionaires can afford to be), willing to risk a few million dollars to put their money where their mouths were in support of sustainable development.

In his two books, *Financing Change* and *Changing Course*, which are still used in many US business and economics courses, Stephan Schmidheiny has written that companies should not have to suffer from the tyranny of the quarterly statement imposed by capital markets that demand growth at all costs. This is certainly true for a private company like SELCO, which has no need to impose anything but annual targets based on realistic expectations. Other values must apply if "sustainability" is to be possible.

Shortsighted fund managers and amateur administrative minions with their MBA degrees, however, have a different view, all left-brained and less visionary. They want their financial projections met *or else*, and none of them understand the concept of "patient capital," nor the values of trust, loyalty, respect, and humility, which our little enterprise was built on. They don't teach those things at the Harvard, Stanford, or Wharton business schools, which turn out some of the most arrogant and self-important "managers" imaginable, whose confidence exceeds their competence (I've met only a couple of exceptions). As SELCO shares and corporate control changed hands, inexperienced MBA fund managers became the company's bane, nearly undoing the vision for a new corporate paradigm that had been supported and invested in by our two enlightened billionaires.

I was partly to blame for this: We never reached the hugely ambitious targets we'd projected in our business plans, despite high annual growth rates, and I was the one who appointed two Harvard MBAs to my board of directors, including a VP from one of the world's largest financial institutions, and agreed to hire a young, wholly unqualified Stanford MBA to run the company when I stepped down from day-to-day operations. Their world, and SELCO's actual business on the ground in the rural areas of the developing world, could never be reconciled, just as the world of finance capitalism can never be reconciled with the market economy, free enterprise, fair trade, and sustainability. As David Korten has written in *The Post Corporate World*, "Finance capitalism makes money from money, without the intervening necessity of engaging in productive activity .... In a market economy, investment is about creating and renewing productive capacity to meet future needs. In a capitalist economy, investment is about making money."

Our narrowly focused bean counters could never understand *Financial Times* writer John Kay's dictum (quoted in Chapter 7), that "the most profitable companies are not the most profit oriented. Individuals who are most successful at making money are not those who are most interested in making money." Unfortunately, some of our directors and shareholders' representatives (not necessarily the shareholders themselves) were *only* interested in making money and could not understand that the foundation of our business was *service*. They could not see that the SELCO companies existed *only* because our hundreds of workers were dedicated to the company's *mission* and often worked without pay when the revenues could not meet the payroll. Mission and service came before greed. This isn't a concept that business schools are willing to teach, despite its truth. Thus, MBAs became the rogue elephants in the cirque du soleil, but SELCO survived them too.

When megacorp executives deride our little company for its slow growth, financial inexperience, managerial deficiencies, and what seems to them to be an irresponsible, steadfast pursuit of an impossible dream, I look at the photo on my bulletin board of Ken Lay, that beacon of corporate savvy, rectitude, and experience, the king of big-growth capitalism ... being led away in handcuffs. It always makes my day.

SELCO survived Enron. It also survived the energy dilettantes at the World Bank and the other bilateral and multilateral development agencies, the universities, the energy academics, and the solar institutes with their plague of renewable energy conferences. But mainly it survived our own ignorance of how to do this. We had no experience as businesspeople or energy entrepreneurs, only some vague idea that solar power would be good for poor people in developing countries who had no electricity and no prospect of ever getting any unless a commercial enterprise could deliver it to them. Only fools rush in to the most difficult and frustrating energy market in the world, notwithstanding that it's also the largest.

But we were happy fools because we all felt very good about how we were seeking to make a living. I would say to anyone out there who may wish to pursue a similar career path: If a bunch of inexperienced young people led by a middle-aged former journalist without specialized academic credentials, a background in business, or any money can build an enterprise like SELCO, almost anyone can. *You just have to want to do it.* In the most difficult times I was often inspired by the words of W.H. Murray, a Scottish Himalayan climber, which I had framed and put on my shelf:

Until one is committed there is hesitancy, the chance to draw back, always ineffectiveness. Concerning all acts of initiative (and creation), there is one elementary truth, the ignorance of which kills countless ideas and splendid plans: that the moment one definitely commits oneself, then Providence moves too. All sorts of things occur to help one that would never have otherwise occurred. A whole stream of events issues from the decision, raising in one's favour all manner of unforeseen incidents and meetings and material assistance, which no man could have dreamt would have come his way.

I also took heart from Goethe's couplet, which had guided me for years:

Whatever you can do, or dream you can, begin it. Boldness has genius, power and magic in it.

Perseverance and persistence always win the day.

CHAPTER X

# The Solar Age For You and Me

At this point, the obvious question is: If 50,000 poor families in the Two
Thirds World can get their electricity from the sun, why can't we? Or,
if one small company like SELCO, plus hundreds of smaller players and
large ones like Shell Solar, can deliver solar power and light as an affordable
option to several million people (averaging five people to a family, plus businesses and institutions) in many of the world's least economically advanced
countries, what is so hard about doing it in North America and Europe?

The answers are: it's not so hard, and we can do it.

Although 25 years have gone by since our Department of Energy–sponsored Gold Lake Conference in Colorado, where Tom Tatum and I tried to
figure out how to sell solar to the American people and failed, the good news
is that it's not too late. In fact, back then it was too early. Idealistically, as
devotees of President Jimmy Carter and his vision, we wanted to help America
escape its dependence on foreign oil — we already knew how risky it was to
rely on the Middle East. But oil remained cheap, and by 1986 it got cheaper
as OPEC drove the price from $30 a barrel down to $10, putting American
domestic oil drillers out of business and nearly wiping out the nascent solar
water-heater industry. Oil stayed cheap — relative to inflation — right up to
the year I started writing this book.

In the spring of 2004, with the Iraq war raging, the dollar falling, China
booming, and OPEC scheming, Tom called me up, Cassandra that he is, to
say, "Fifty-dollar-a-barrel oil by August." He was wrong ... by a week. It hit

$50 the first week of September. By mid-2005 it had hit $60. It's not clear where it will be when you read this, but it will never be $30 again.

The cost of oil, of course, doesn't directly affect electricity prices; it affects transportation. But since everyone in North America drives, everyone knows "energy" is getting more expensive. And good old supply and demand began driving up the price of natural gas significantly. That hits people at home: cooking, heating, hot water. Heating oil also increased in price in 2004.

And then a debate that had been going on for years among the world's oil analysts went public, thanks to a spate of new books, articles, and massive Internet coverage about *peak oil*.

What does this mean? In short, oil production "peaks" when you extract more oil than ongoing exploration can discover in a particular market or country or oil field. US oil production peaked in 1970, and in many other oil-producing countries it peaked not long after that. In 2004 it became clear oil may peak globally much sooner than anyone realized. What we don't know is the extent of the "known" reserves in Saudi Arabia, where oil is expected to peak last — between 2006 and 2015 according to Richard Heinberg, author of *The Party's Over* (2004), one of the most thorough and insightful books on the end of the age of oil. When the demand for oil out-strips our capacity to produce it, the party will indeed be over. No, we won't run out of oil right away, but for the first time oil will go from being a buyer's market to a seller's market forever after.

In *The End of Oil* (2004), another excellent book on the same frightful topic, Paul Roberts writes: "As we approach the peak in production, soaring prices — seventy, eighty, even a hundred dollars a barrel — will encourage oil companies to scour the planet for oil." He predicts they will be unable to stave off disaster or even soften the landing, and then he demolishes the arguments of ill-informed business leaders and ostrich governments who think our reserves are enormous. "In short," he writes," oil depletion is arguably the most serious crisis ever to face industrial society."

Vice President Richard Cheney, no dummy, knows this. The oil men, George W. Bush and Dick Cheney, and woman, Condoleeza Rice, also from the oil business, who run America know exactly why the United States is in Iraq: *to build permanent military bases to ensure the oil keeps flowing from Saudi Arabia, and from Iraq, in our direction no matter what happens, no matter what it costs in lives or treasure.* If they really believe they can build a democracy in Iraq, they are dumber than I thought. I don't think they even believed their own lie that Saddam was connected to 9/11: I think they believe the American people are stupid enough to believe anything they tell them. So far, they appear to be right.

I believe they think the American people would panic if they knew just how vulnerable our supply of oil is, and they think Americans are not as likely to panic over a patriotic war wrapped in red, white, and blue ... and a lot of yellow-ribbon magnets on SUVs that say "Support our Troops." And when self-entitled Hummer drivers can't get gas, there *will* be panic. The last to know always take it the hardest. Unfortunately, there are a lot more SUV owners than hybrid Prius owners (but 2004 saw the introduction of the first hybrid SUVs, so things are changing, and Toyota sold 11,500 Prius cars in just one month — March 2005).

In *Out Of Gas*, his short but devastating history of fossil fuels and electricity published in 2004, Caltech professor of physics David Goodstein predicts doom and gloom. Goodstein writes, "Our way of life, firmly rooted in the myth of an endless supply of cheap oil, is about to come to an end .... When the rate of increase of known reserves reaches zero (which for all practical purposes may already have happened), we will for the first time in history be consuming oil faster than we are finding it." This point of peak oil, described earlier, is known as "Hubbert's peak" for the geologist who predicted it many years ago. Goodstein continues: "The Hubbert's peak assumption is that the crisis will occur not when the last drop is pumped, but at the halfway point, where falling supply meets continued rising demand. *If we have already consumed nearly half the oil there ever was, the crisis can't be far off"* (italics mine). He concludes that if we don't turn to the sun for energy, "civilization as we know it will come to an end sometime in this century unless we can find a way to live without fossil fuels."

Got that?

I think people are getting it. Okay, maybe not the oblivious Hummer owners, but lots of people. For the first time in 25 years I meet people everywhere who ask me, "How can we get a solar system for our house?"

Why do they want to do that? Because they are worried. Four hurricanes in Florida in 2004 knocked out power to millions; another million were without power for weeks in Maryland and Virginia thanks to Hurricane Isabel in 2003. I've already mentioned the northeastern blackout of 2004 that plunged 50 million people into darkness. And the memory of the California blackouts in 2001 and 2002 are still fresh. These things get on people's nerves. Even though it's electrical power and not gas or oil, they wonder if the weather could be related to global warming and hence to fossil fuels.

Despite this worry and despite all the signs pointing in the opposite direction, a December 2004 energy study by the National Commission on Energy Policy, chaired by William K. Reilly, former head of the Environmental Protection Agency, and funded by the Hewlett Foundation, said we should use *more* coal! Which means we will, which could mean more unpredictable climate change, more severe weather, tornadoes, hurricanes, and typhoons, which have a tendency to destroy electricity grids. Just ask the people of Haiti or of the island of Grenada, where it will be years before they see power lines again, if they ever do. The $5-million commission study, "Ending the Energy Stalemate," offers a cocktail of solutions to deal with both energy shortages and global emissions, but it scarcely considers solar, which angered the membership of the Solar Energy Industries Association <www.seia.org>, who chafed at the report's focus on so-called clean coal. It does, however, propose more energy efficiency and conservation and recommends $300 million for solarelectric R&D. Meanwhile, the US secretary of energy predicts that to meet America's demand for electricity over the next 20 years, we need to build 1,300 to 1,900 fossil-fueled power plants. What will those do for the skies over Yosemite or the Shenandoah?

The National Commission on Energy Policy report is certainly *not* the "blueprint for a solar economy." That kind of visionary thinking doesn't have much traction in today's energy world with so many competing energy-producing interests. *The Blueprint for a Solar America* was published in 1978 by the Solar Lobby — to no avail. *Energy Future: The Report of the Energy Project at the Harvard Business School* also outlined what "a solar America" might look like and proposed a sensible, achievable energy mix in response to "hostile oil" from "Saudi Arabia, a shaky foundation for Western Civilization." But you probably have not seen that study around lately, even though the *Wall Street Journal* praised it as "Heroic .... A truly magnificent book." It was published in 1979. Nor have you seen the popular paperback *The Coming Age of Solar Energy*, a remarkable story of "the history, the technology, and the future of sun power," because it was published in 1975.

So what does all this have to do with solar energy for you and me?

I'll get to that shortly. I've tried hard not to write a "should do" or "must do" energy or environmental policy book, and have eschewed "how to," since

you can find that information else-where (see Bibliography). This is meant as a "can do" book. You *can* do something ... something simple and practical that will save you money, guarantee a supply of electricity, and make you feel good (priceless). I mean, if 50,000 peasant farmers no longer have to worry about getting kerosene for their lamps or getting their electricity supply from fossil-fueled power plants ....

Shell Solar

Meanwhile, bear with me as I try to frame the endless energy debate as it stands in the second half of the first decade of what *will* be, for lack of any other choice, the solar century.

California suburban home with PV system mounted on tile roof.

In 2004, solar power went mainstream. According to Rona Fried, president of Sustainablebusiness.com, "2004 may well be remembered as the watershed year for solar energy. It's been a long time coming, but in this 50ᵗʰ anniversary of the solar cell, the industry is showing clear signs it is indeed at the tipping point toward mass commercialization." I was beginning to wonder if I'd ever see this happen! We heard bipartisan voices resoundingly supportive of moving beyond carbon fuels. In December 2004 I attended the third conference of the American Council on Renewable Energy (ACORE), which is part of the World Conference on Renewable Energy (WCRE). I had been at the founding meeting with Hermann Scheer and Michael Eckhart in Berlin in 2000 and never thought anything would come of this, yet *another* solar-energy lobbying and policy group. But thanks to Mike, one of the world's most enthusiastic and tireless energy activists, whom I met when we hired him to advise on how to launch SELCO in 1996, ACORE has managed to push the levers of power like no other organization before it. ACORE was a perfect counter to Dick Cheney's National Energy Policy Development Group, which formulated its plans in secret (secrecy upheld by the US Supreme Court). This private task force of energy corporation executives under the vice president's chairmanship stated: "America in 2001 faces the most serious energy shortage since the oil embargoes of the 1970s." Scary stuff; scary response. Then came 9/11.

In *Crossing the Rubicon: The Decline of the American Empire at the End of the Age of Oil* (2004), likely the most important book yet written linking oil, energy supply, national security, and US foreign policy, author Michael Ruppert writes: "The United States has chosen to address the problem of Peak Oil in the most brutal, venal, and shortsighted way available: by using military force to commandeer what remains of the world's rapidly vanishing fossil fuels." In an exhaustively researched 600 pages, the former Los Angeles police detective examines the unthinkable, which I shall not attempt to endorse or summarize here, other than to say that if you want to continue to sleep well, do not read this book. Ruppert makes it very clear that the world's superpower is under the leadership of a man who wants the energy-guzzling

Texas way of life for all, since that's all he knows. This is the same man who has said, "The American way of life is not negotiable," and who will preemptively invade any country that gets in his, or our, way. Ruppert connects the dots like no one else.

If what he says is true, then 2004 was also a tipping point for danger, despair, and doom in view of the fact that the "Bush/Cheney Junta," as Gore Vidal calls this administration, was reelected in November. Government lying on a scale not seen before in the United States got results.

But back to the ACORE leadership conference and the 500 attendees at the magnificent Cannon Caucus Room of the US House of Representatives on Capitol Hill. A *Republican* senator, Colorado's Wayne Allard, kicked off the meeting, stressing the importance of using renewables in light of the energy security crisis, noting that renewable energy is now a $22-billion industry, and mentioning the solar-powered cabin he's building in the mountains. I remembered his colleague, Trent Lott, Senate majority leader, publicly dismissing solar power as "hippie technology."

Allard was followed by Admiral James Woolsey, former director of the CIA, who said, "If you drive an SUV, you may be contributing to the sinking of Bangladesh." He drives a Prius and had just put a solarelectric and solar heating system on his house. Woolsey said we had no choice but to use alternative energy sources because we were going to be at war with militant Islam for decades — and it is a war we may not be able to win. It was clear he and fellow speaker Robert "Bud" McFarlane, President Ronald Reagan's national security advisor, were concerned that Bush's invasion of Iraq was not helping to win this war but instead had unleashed the whirlwind. Security expert Frank Gaffney implied that in "Islamic fascism" we faced a threat worse than Nazism or communism, so we'd better damn well get serious about finding alternatives to the remaining oil that Islamists control or will soon control thanks to Bush and Cheney's calamitous, unwinnable war.

Next came the head of GE Energy, who noted that the world's largest product manufacturer now viewed wind power as a power source as conventional as gas turbines. "GE is betting on a greener future," he said. In 2004,

GE purchased Astropower in Newark, Delaware, the largest American-owned manufacturer of solar photovoltaics, which it intends to turn into a billion-dollar business.

"Edison would be pleased," I told him afterward.

"He's our founder," he said proudly.

"I know," I said, reminding him of Edison's quote in favor of using the sun's energy: "What a source of power!"

While we inelegantly munched our brown bag lunches catered by the Republican Club of Capitol Hill, Amory Lovins announced the results of his new study, commissioned by the Pentagon, "Winning the Oil Endgame." Winning that game, he asserted, will be the "cornerstone of the next industrial revolution ... and, surprisingly, it will cost less to *displace* all of the oil that the U.S. now uses than it will cost to *buy* that oil." Ever the dreamer, Lovins nonetheless is right, about as right as President Carter was in 1979 when he said, "We must end our dependence on foreign oil, and we must get 20 percent of our energy from the sun by the end of the century." He was speaking of the *last* century. So it goes.

Historical perspective was provided by Jay Hakes, director of the Carter Presidential Library, who showed a video of one of Carter's visionary solar speeches of 25 years ago, astonishing the younger members of the audience who thought solar was something new. Hakes reminded us of what President Richard Nixon said in 1973: "Let us set as our national goal, in the spirit of Apollo, with the determination of the Manhattan Project, that by the end of this decade we will have developed the potential to meet our own energy needs without depending on any foreign energy source." Amory Lovins and Richard Nixon as co-equal visionaries!

The greatest solar visionary of our age, Dr. Hermann Scheer, a longtime member of the German Bundestag, author of *A Solar Manifesto* (1991) and *The Solar Economy* (2002), gave a keynote talk at the ACORE forum. He explained how we could "break the globalized chains of the fossil fuel supply networks," which have tied down the global economy for so long, running it like a private cartel. How we do this is all there in *The Solar Economy*, the most important, complete, and thoughtfully reasoned argument for a global solar economy yet published (and nearly impossible to buy in the United States).

I first met Hermann in 1991, at a solar PV conference in Zimbabwe that he organized on behalf of the European Union. I was there designing the UNDP's first national solar rural-electrification project. While he'd relentlessly pushed the EU to fund solar for developing countries, his real success came at home as the "father" of the German national solar photovoltaics program, which bypassed government, industry, and the vaunted "market" to institutionalize the use of PV *by the people* ... who voted to buy it, to use it, to pay for it, and to guarantee its growth as a mainstream energy business. We can only dream of this in the United States, or write policy papers and books and hold conferences; in Germany they did it, they are *doing* it. Germans are reportedly installing 400 SHSs a day.

"We didn't wait for the market or for a consensus or for the conventional energy system to do the job," he told me later. "There is no free market in energy, and governments are the last to move. So we relied on the people. We went to the people with the initiatives who voted in favor of solar electricity. But we are in a race against time to replace conventional energy with renewables. What we need now are practical protagonists."

"Practical" has never been in the lexicon of energy analysts and theorists, but nothing under the sun is more practical than PV. I think the Germans are among the first to know it, even though it was invented in the United States.

Have I mentioned the environment? Jonathan Lash, president of the World Resources Institute, and Chris Flavin, president of the World Watch Institute, both spoke at the ACORE conference, emphasizing not just the well-known environmental consequences of continuing to rely on carbon fuels, but offering examples of positive approaches being undertaken by industry and by state and local governments. Flavin, however, pointed out that Germany is first in wind power, China first in solar heating, Japan first in PV,

Brazil first in biofuels — and the United States is last in everything but oil consumption. Washington's environmental nonprofit organizations have been leaders in promoting renewable energy over the past decade; Greenpeace put solar modules on its roof downtown.

The tipping point year of 2004 was the fourth hottest year on record — nine of the ten hottest have occurred since 1995. It boasted the warmest October ever and a record number of tornadoes (1,734) in the United States. It was the costliest year for the insurance industry, which faced $36 billion in claims, largely due to the most weather-related disasters ever to occur in a single year. In 2004, *Boiling Point* was published, written by Pulitzer Prize-winning author Ross Gelbspan. If you are a fan of horror movies, forget them and just read this book; if you don't like to be scared, skip it. Besides reminding us how serious a threat climate change is to our species, Gelbspan takes issue with nearly everyone, as evinced in the book's subtitle: "How Politicians, Big Oil and Coal, Journalists, and Activists Have Fueled the Climate Crisis — and What We Can Do To Avert Disaster." He castigates the Washington energy and environmental policy community, which he says has been hijacked by the fossil-fuel lobby, and writes: "The solution to the climate crisis involves a high-stakes battle with big coal, big oil, and the immense financial resources and political levers at their disposal. Soft approaches do not normally prevail in hardball competition."

He offers a Marshall Plan approach to "rewiring the world with clean energy" and says, "A properly funded global transition to clean energy would create millions of jobs in poor countries and substantially raise the living standards in the developing world." Like Scheer, he proposes a new industrial revolution ... nothing short of an "energy revolution." He slams those who think a market-based approach will move us to clean energy; also like Scheer, he believes people will have to *lead* so that the market, and lastly the government, will be forced to follow. People as leaders — what a radical democratic idea!

In December 2004, 6,000 government leaders, delegates, environmentalists, energy activists, and executives gathered in Buenos Aires to discuss the Kyoto Protocol, ratified by 130 nations, but not by the United States. A

Bush appointee, US delegate Paula Dobriansky, told the meeting with a straight face that science cannot say "what constitutes a dangerous level of warming," implying no action was needed. Opposed to the idea of "global warming," the Bush people also opposed the term "climate change." But even without the US, the Kyoto pact will begin to generate new policies, practical solutions, economic incentives, and popular action worldwide.

Also in 2004, the tipping point year:

+ Former president Bill Clinton, in a landmark speech at New York University, said we should stop "bellyaching and whining" about political obstacles and undertake a new effort to address the "intertwined problems of energy dependence and global warming."

+ Crude oil broke the psychological barrier of $50 per barrel and passed $55 before falling back — temporarily.

+ GE entered the solarelectric product manufacturing business, and analysts said GE, now in wind *and* solar, stood to become the world's largest alternative-energy company by the end of the decade. (In a May 2005 *Washington Post*, GE's CEO Jeff Immelt and World Resources Institute's Jonathan Lash wrote, "[The United States'] primary objective must be to revolutionize the way we produce and consume energy. Diminishing gas reserves, continued reliance on foreign and sometimes unstable energy sources, and global climate concerns demand nothing less.")

+ GM and DaimlerChrysler teamed up to develop a new hybrid vehicle-propulsion system, even though the president of GM had said only a year earlier that there was no market for the Prius.

+ Bob Stempel, ex-CEO of General Motors, now head of ECD Ovonics, manufacturers of UniSolar PV modules, told *Fortune* magazine that "solar power ... is becoming mainstream ... [and] is growing at 25% a year [and] the business case for solar is becoming clearer .... Solar can become a free source of power for you."

- Americans bought 80,000 hybrid vehicles.

- The touted "hydrogen economy" was finally seen as the public relations and financial investment scam it is, even while GM continued to promote it. (It is easier to promote a futuristic technology than to actually build electric cars, which GM ceased doing.) At the ACORE meeting, not one speaker even mentioned hydrogen or fuel cells.

- Fifty-three percent of Coloradans voted for a state Renewable Energy Portfolio Standard (while also voting for George Bush), which requires major utilities to buy or produce 10 percent of their power from renewable energy sources by 2015, while establishing net metering standards for grid-connected household PV systems.

- California's Governor Arnold Schwarzenegger proposed a statewide subsidy for solar PV systems, expanding the successful solar power incentives pioneered by S. David Freeman and Angelina Galiteva at LADWP, America's largest municipal utility.

- New Jersey enacted the biggest state subsidies for commercial solar PV installations in the nation, and the New Jersey Board of Public Utilities approved a fund of $745 million for renewable energy and energy efficiency.

- The Pennsylvania legislature passed its Advanced Energy Portfolio, aimed at producing 408 mW of solarelectric generating capacity by 2015 — enough power for 300,000 homes.

- Florida Power and Light launched "Sunshinenergy," quickly signing up 10,000 customers for an additional $9.75 a month, proving people are willing to pay extra for clean energy. For each 10,000 volunteers, FPL will build a 150 kW solar PV facility to feed solar power into the grid.

- In Washington, DC, the city council voted to require DC utilities to obtain 11 percent of their electricity from renewable energy by

2022. "There's plenty of sun in DC," said Rhone Resch, president of the US Solar Energy Industries Association. "Every rooftop in the city has the potential to generate its own electricity."

• *New York Times* columnist Thomas Friedman proposed George W. Bush cement his name in history by embarking on "our generation's moon shot: a crash science initiative for alternative energy and conservation to make America energy-independent in 10 years." (Fat chance, but it's the thought that counts.) Two years earlier, always the dreamer, Friedman had proposed that the president call on "every U.S. school to raise money to buy solar-powered light bulbs for every village in Africa that didn't have electricity so African kids could read at night."

• For the first time, *National Geographic* magazine ran a cover story on global warming, "Bulletins from a Warmer World," citing the melting of Alaska's permafrost and of glaciers worldwide. A few months earlier its cover story was "The End of Cheap Oil."

• *Fortune* magazine featured a cover story about "Kicking the Oil Habit."

• In a poll, 82 percent of Californians said they would support the state doubling its use of renewable energy, already at 12 percent without hydro.

• At the Renewables 04 Conference in Bonn, Germany, Chancellor Gerhard Schroeder told 3,600 participants from 154 countries that we faced a "nightmare scenario ... that can be avoided only if we radically reduce greenhouse gas emissions." (Can you imagine a US president speaking at a renewable energy conference?)

• Osama bin Laden — still at large — videotaped a statement expressing his desire to see Saudi and Iraqi oil priced at $100 a barrel (given the $5 billion-a-month cost of the Iraq war, Iraqi oil is already costing Americans about $10,000 a barrel).

- The sales of GM's outrageously outsized Hummer collapsed.

- The worst environmental president in American history was reelected, scaring the bejeesus out of every concerned American. He appointed a man with zero energy experience as secretary of energy. The president's popularity belied the seriousness with which more people than ever before viewed the energy and environmental crisis.

What does all this have to do with you and me? A lot. What I've tried to show thus far in this chapter is that in 2004 the environmental, energy, economic, psychological, cultural, and informational "climate" — and the *actual* climate — changed as they never had before. Energy costs, energy security, and clean energy became the popular mantras of the day as literally billions

Sharp Corp.

All-solar house with architecturally designed PV system.

of people became concerned about at least one of these issues, if not all of them. Change doesn't come about because of policy studies, books, reports, articles, strategy meetings, and conferences, although these help; change comes because people want it. For the first time in 25 years of watching the energy crisis unfold, I believe people everywhere are now ready to do something about it. In fact, people *are* doing something, as you'll see shortly.

As Caltech's David Goodstein wrote in 2004 in the *Los Angeles Times*, "For as long as we have the sun, we have at our disposal a steady stream of energy amounting to about 300 watts per square meter averaged over the face of the Earth." Solar power will enable us to win "the real war," he says. "The alternative is to go on hunting terrorists while our civilization slides into oblivion." Also in 2004, Paul Roberts, writing about "the undeclared oil war" in the *New York Times*, added, "This only hints at the energy crisis facing the developing world, where nearly two billion people — a third of the world's population — have almost no access to electricity ... and thus are condemned to a medieval existence that breeds despair, resentment, and ultimately, conflict."

These concerns are driving the boom in solar power in the West, just as the wholesale lack of electricity in much of the Two Thirds World contributed to PV's growth over the past 15 years. In 1997, three-fifths of all PV cells ended up in the developing world, mostly to supply rural markets and telecommunications. By 2004, shipments to developing countries from the West were around 10 percent of production. In fact, PV module manufacturers in India, Thailand, and Mexico were shipping the bulk of their PV modules to Europe and Japan, while still supplying their own markets. The best thing about the growth of solar electricity in the West is that it is not a victim of, nor dependent on, world development agencies like the World Bank, GEF, IFC, USAID, UNDP, and bilateral donor agencies that have hampered as much as they have advanced the use of solar power in the developing world (as I've tirelessly pointed out in earlier chapters). In fact, solar power is booming in America, where successive federal administrations have deliberately chosen to stay out of it. *People* are making it happen, and the market economy is responding; free enterprise is leading the way, from the multinationals to the smallest garage-based entrepreneur. Many state and local governments are responding with packages of incentives, tax credits, and direct subsidies. For example, in 2001, San Franciscans voted for a $100-million bond issue to finance solar power facilities (see <www.dsireusa.org> and <http://forsolar.org>).

In its November/December 2004 issue, *Mother Jones* magazine published the most important breakthrough article on solar electricity I have ever seen (and I have file cabinets full of 25 years' worth of press clips on solar energy). Despite being a leading progressive journal (named for the famed labor organizer of the early 20th century), *Mother Jones* has always been skeptical of solar energy, and I can't recall it ever publishing a useful or accurate story on the state of the industry. But this article, "One Roof At A Time," by environmental journalist and author Bill McKibben, reports on how "solar power is edging into the mainstream." I had been so focused for so long on bringing solar electricity to villages in the Two Thirds World that I nearly missed the story of how the One Third World was going solar.

McKibben, who put 12 PV modules on his own house, interviews happy PV users around the country who *love* watching their electric meters run backward on their "grid-tied" systems as the solar power they don't use goes out over their grid connection into the local utility's system. Many become energy-efficiency enthusiasts so they can make the meter run backward faster. McKibben notes that most months his electric bill gets "substantially smaller" as he earns credit from the utility (in some states the "net metering" law allows credit to be rolled over month to month, while in others it has to be used each month, like most cellphone minutes). McKibben, also a skeptic until he installed his own system, reports that "more solar power has been harnessed on the world's rooftops in the last two years than in all of previous history."

He cites the Japanese program, in which over 150,000 grid-tied rooftop PV systems have been installed, and the German program, based on Hermann Scheer's "feed-in tariff" program that forces utilities to buy home-made solar power at a generous price, which is growing at 50 percent a year, with 100,000 rooftops covered in PV thus far (the aim is to have a million roofs). McKibben reports that global installed capacity will be a gigawatt in

2004, while in the United States the domestic PV market is growing at 60 percent a year. A big chunk of this is in California, where "within a decade the state could have more solar panels than any single nation," according to McKibben.

In December 2004, Governor Schwarzenegger's deputy secretary for energy, Joe Desmond, announced that California could have *one million homes and buildings producing solar energy by 2018*, with half of all new homes powered by the sun. The goal is to pass the Million Solar Roofs initiative, which would ultimately lead to the generation of 3,000 mW of power from the sun within 13 years (one mW will power 750 houses). It would require builders to offer solar power as an option and would implement other guidelines echoing past requirements to install low-flush toilets, code insulation, and energy-efficient appliances. The state would also add a small charge to utility rates to fund the solar subsidy of approximately $2,800 per kilowatt installed. As already noted, 82 percent of Californians have indicated they support doubling the state's use of renewable energy. The people will decide.

BP Solar

BP Solar installer assembles grid-tied rooftop PV system in the UK.

Costs have always been the issue, but now people recognize the greater cost of local air pollution, energy security, and global climate change. At the same time, costs of PV have fallen dramatically, thanks to expanded production serving growing markets. Homeowners have figured that paying 15 cents per kWh more for solar than for utility power is worth it because they can reduce their electric bill and calculate full payback in a set time period (five to ten years) based on the size of their SHS. It's also worth it because they feel good and know they are doing something about the environment. For those who put personal security ahead of society's safety, they know that buying a solar system with backup batteries will provide them with power in the event of outages caused by humans or by acts of God.

However, in the United States, the biggest PV market is still commercial: institutions, businesses, and industries are buying large rooftop installations. Companies like Ford Motor Co., PepsiCo, Staples, Whole Foods, Lowe's, FedEx, Toyota, and Johnson & Johnson have employed large-scale rooftop PV systems at various facilities for the double reason: energy savings in the long term and "good citizen" PR points immediately. More and more companies believe going green *gets* the green. Additional corporate buyers of PV are IBM, Dow, DuPont, Alcoa, Intel, HP, Pitney Bowes, Kinkos, and General Motors.

Government agencies are buying large to very large PV systems (several hundred kilowatts to above a megawatt), many installed by Berkeley's Powerlight Corporation, including the Santa Rita Jail in Alameda County, the Solano County Government Center, the City of Santa Monica (to charge electric vehicles), the Anaheim Convention Center, the US Postal Service, the US Coastguard headquarters in Boston, the Coronado Naval Base headquarters, US Park Service ranger stations, the student union at University of California at Berkeley, and all over the Cal State campus at Hayward. Resorts like the Mauna Lani in Hawaii have covered their roofs with PV, as have many California wineries. This list is just a tiny preview of America's thousands of commercial PV installations. (For more information on what one small company has done, go to <www.powerlight.com>.)

The Pentagon is now surrounded by a forest of PV on poles, which light the parking lots; the White House maintenance annex has a PV system on its roof; and the United States Mission to the United Nations in Geneva, Switzerland, was recently solarized. In Washington, DC, what was once the world's largest "building integrated" rooftop PV array, at Georgetown University, is still steadily producing power after nearly 25 years, but today such 300 kW solar systems are commonplace around the country. BP Solar, which in 2004 was profitable for the first time in 25 years, is covering hundreds of BP's gas stations here and in Europe with its own PV panels — ironically, to power the pumps that deliver fossil fuel. You don't read about these installations in the national press, but go online to <www.renewableen-ergyaccess.com> or <www.solarbuzz.com> and then do a little follow-up research on the Web, and you'll be amazed to see that solar has become a big business, and that "big business" is putting it to work.

And the big players have only gotten bigger: Sharp, in Japan, with a 400 mW production capacity, is the largest manufacturer of photovoltaics in the world. Shell Solar, which sold over 100 mW in 2004, is next. Third is Japan's ceramic and semiconductor giant Kyocera; followed by British-owned, US-based BP Solar; RWE, Solarworld, and Q-Cells in Germany; and then Mitsubishi, Spain's Isofoton, and Sanyo. Dozens of smaller manufacturers are in the game worldwide, including the few American PV makers: Evergreen Solar and Konarka of Massachusetts, Energy Conversion Devices (United Solar) of Michigan, and SunPower Corporation of California. Hundreds of smaller firms around the world manufacture silicon wafers from solar-grade silicon, the building block of PV. At this time it's not clear where GE Solar stands nor where it is going, but the former Astropower plant in Delaware is shipping product. GE could climb to near the top of the PV manufacturers' list by the time you read this.

In their book *The Oil Factor: Protect Yourself and Profit from the Coming Energy Crisis* (2004), Stephen and Donna Leeb explain that you can make money off the oil crisis ... by investing in alternative energy companies: "If the U.S. can establish a leadership position in the development of alternative energies, we'll not only have successfully met the tremendous challenge of declining oil supplies. We'll have built a vital new industry based on vital new technologies that will spur economic growth and ensure our continued hegemony in the world community."

Not to be overlooked are public and investor-owned utilities — like the Sacramento Municipal Utility District, Florida Power & Light, Pacific Gas & Electric, and Austin Energy in Texas — which buy huge quantities of PV and set up large, central, solar power arrays to feed their grids. There are many more, too numerous to mention here, as well as independent energy contractors like GreenMountain, which finance solar arrays and large roof installations and sell their clean power directly to ratepayers.

How big solar PV has become was highlighted by CLSA Asia-Pacific, a large Hong Kong-based global investment bank and brokerage. In 2004 it released its "Solar Power Sector Outlook." The analysts wrote, "Initially skeptical, we have become enthusiastic about solar power because it has realistic prospects for revenue to expand from U.S. $7 billion to U.S. $30 billon by 2010." The exhaustive 64-page industry analysis pronounced that the "solar power industry is now profitable." Capital markets are responding to investment opportunities in solar power like never before, and in 2005 the private sector ramped up module and cell production to meet demand.

"In 2004, worldwide photovoltaics production surpassed one thousand megawatts [a gigawatt!] for the first time — another thirty percent plus year," Paul Maycock reported in February 2005. Paul, a director of SELF and the world's foremost authority on the PV industry, is forever talking about the "booming" PV business, which he analyzes in his annual *World PV Market Report* (see <www. pvenergy.com>).

A September 2004 *Business Week* article, "Another Dawn for Solar Power," focused on "tech breakthroughs," which it reported had the prospects to turn both commercial and residential rooftops into power plants to "compete with fossil fuel-generators in markets where electricity costs at least 10 cents per kWh." Daytime peak power in many places costs well above that. But it isn't technology "breakthroughs" that are now delivering solar power to tens of thousands of American users; it is the marketing and sale of existing work-horse solarelectric technologies by relatively small operators — solar dealers and installers — all over the globe and in every state in America. Thanks to them, and no thanks to governments that subsidize the nuclear and fossil-fuel energy supply system to the tune of an estimated $300 billion per year, the solar revolution we plotted 25 years ago is finally getting organized.

Here is where *you* come in. If the solar age is to be about you and me, we need to be the customers for solar power. It's here; you just need to look for it. It's affordable; you just have to prioritize it as something you need and want. Hundreds of small installer/dealers out there are as busy as they can be installing mostly grid-tied residential PV systems all over the United States. They are largely "mom and pop" operations — mostly "pops," although there is a substantial number of "moms" and single women working as PV technicians and installers. Most installers are state-licensed electricians with PV certification provided by the North American Board of Certified Energy Practitioners following a rigorous training and testing procedure. The 20-year-old Florida Solar Energy Center is also a trainer of PV technicians and has developed its own SHS standards <www.fsec.ucf.edu>.

The solarelectric retail and distribution business in North America today is in the hands of small players, not the large corporations mentioned above (with the exception of BP Solar, which is now marketing its SHSS through Home Depot outlets in California and Costco stores elsewhere). Some of these, like Xantrex, Sunwize, Outback, and Northern Power Systems, also manufacture electronics and other components. Others, like Genself, Solar Depot, Energy Outfitters, Global Resource Options, Renewable Technologies, PV Powered, Solar Design Associates, Solar Outdoor Lighting, Dankoff Solar, Real Goods, Schott Solar, and time-warped but proficient Backwoods Solar Systems, specialize in retailing residential solar-electric systems, mostly on a custom house-by-house basis. Some supply subdivision building contractors who are preinstalling SHSs in their new homes or who are offering solarelectric systems as an add-on or upgrade. At the same time, hundreds of small PV resellers, practitioners, and installers dot the landscape.

All these enterprises and many more are online or can be found through the industry bible, *Home Power Magazine* <www.homepowermag.com>. Many have local associate installer/dealers around the country. Another key publication is *The Real Goods Solar Living Sourcebook* edited by John Schaeffer and Douglas M. Pratt. You can also pick up the *Consumer Guide to Solar Energy* (2002) by Scott Sklar and Kenneth Sheinkopf, renewable energy veterans who cover not just PV, but also solar hot water, pool heating, solar cooling, and space heating. Especially for Europeans, but worth a visit by anyone, is the PV business website <www.solarplaza.com>. Additionally, there is the excellent website of the globally oriented European Photovoltaic Industry Association <www.epia.org>. For Americans there is <www.cleanedge.com>, a business guide to renewables in the United States. And don't miss a visit, in person or on the web, to the Solar Living Institute in Hopland, California <www.solarliving.org>.

If illiterate farmers in India can figure out that solar is better for them than the alternatives — kerosene and darkness — why can't we make a similar life-changing decision to install an affordable energy system in our homes that will reduce greenhouse gas emissions by the amount of conventional power we offset — a simple, affordable consumer appliance that can also provide electricity during a power blackout? We can, and we are. (Well, I'm not, yet, because I have to get a waiver on the covenants in our "traditional neighborhood development," the new-urbanist community of Kentlands, Maryland. I'm working on it, now that I'm permanently home from my far-flung solar adventures.)

I firmly believe in the old capitalist adage, "Supply creates its own demand." SELCO proved that by reaching out to its 50,000 + solar customers. As more and more people hear about and see solar at work in the United States, and as channels of supply become more apparent and accessible, demand will mushroom. It will be we, the people, who create the new world of clean, independent power, and Americans and Europeans will do it for the same reason that poor farmers buy a solar home system in South Karnataka, India, or Wellawaya, Sri Lanka, or Binh Phuoc Province, Vietnam. Because they can. Even if solar retailers are still hard to find, you can find them; they're out there.

Listen to David Hollister, a solar entrepreneur in Asheville, NC, and owner of Sundance Power Systems, who says, "People are choosing a different path, becoming conscious of their true natures and aligning with higher values that guide them into a more peaceful and sustainable future in harmony with the planet we call home. Through this journey we run face to face into our desires and attachments, into 'our way of life.' We need to make choices that reflect our newfound identities and values." These are the ethics of the solar business.

If you are young and don't have a house to retrofit, another opportunity, promoted heavily by the Apollo Alliance in Washington, is the chance to work in solar PV. A solar retail industry will create lots of jobs. The Apollo Alliance, <www.apolloalliance.org>, believes "freedom from foreign oil" can ultimately produce three million new jobs in the renewable energy industry. But there is no need to wait for Washington policy wonks to have an effect; any young person can learn the trade, get licensed, and jump into a business that is going into overdrive. In the movie "The Graduate," the future was described in a word, "plastics." Now it's "PV," so get on board! Electrical contractors can also add PV specialists and a solarelectric line of products and services. In California there is a shortage of solar technicians. Electrical utility workers have to climb poles, but with solar there are no poles to climb or wires to string, only rooftops to negotiate. (Or one can install backyard pole-mounted solar tracking arrays if a roof is unsuitable.)

Certified PV technician training courses are available year-round from Solar Energy International (SEI) in Carbondale, CO, and other venues <www.solarenergy.org>. Headed up by master trainer Johnny Weiss, SEI has been teaching solar electricity to the old and the young, men and women,

Americans and foreigners, for over 20 years. SEI also publishes the comprehensive *Photovoltaics Design and Installation Manual* for do-it-yourself types.

Do you need inspiration? There will be lots at the second Solar Decathlon on the Mall in Washington, DC, in October 2005, sponsored by the US Department of Energy and BP Solar. A veritable suburban tract of 20 all-solar electric houses, built by engineering students at their respective universities,

Neville Williams

Several of the 16 solar houses built by university students on the Mall in Washington for the Solar Decathlon, 2002.

will be disassembled, shipped to Washington, then reassembled right in front of the US Capitol and opened to the public. It's a sight to behold. (I highlighted the first Solar Decathlon in Chapter 1.)

I'd better mention the two energy bogeymen here, nuclear and hydrogen, since these technologies always enter the energy agora. "Why should we go solar if safe nukes and clean hydrogen are just around the corner?" well-meaning people ask. I'm not a policy writer, so I'll make this quick and quote the experts, including our old pal S. David Freeman, who said this about nukes at a "power lunch" hosted by the Sierra Club a couple of years ago: "It's dead, except in the hearts and minds of the religious believers in nuclear power. After September 11, we are surely not so dumb as to build more Trojan horses in our country. The danger of a penetration into a nuclear reactor — which is difficult but not impossible — is so horrendous that we've got to be out of our minds to build more nuclear power plants. And I say this as someone who's had as much experience with nuclear power as anyone in this country. I shut down eight reactors when I was the head of the Tennessee Valley Authority, buried one at Rancho Seco, and nursed one back to health in New York. But in this age of terror, we just can't have them."

At the roundtable discussion, which included Lord John Browne of BP, Lester Brown of the Earth Policy Institute, architect William McDonough, and Carl Pope of the Sierra Club, Freeman weighed in on hydrogen as well: "The Bush administration has endorsed the hydrogen fuel cell, but there's no program for development of hydrogen fuel. Let's have one seat at the table for common sense, which suggests that clean technology needs a clean fuel to go with it." And with the growing realization that hydrogen is not an energy source but an energy carrier (it always was, but hope got in the way of under-standing), there is the failure of markets to support "the hydrogen economy." By 2002 the hydrogen bubble had burst. One hydrogen buff lamented, as the hydrogen mania wound down in 2004, "I'm afraid that when we finally get people to stop associating hydrogen with bombs and the Hindenburg explo-sion, the next word they'll think of will be scam." Or as James Howard Kunstler wrote in *Rolling Stone* in April 2005, "The hydrogen economy — widely touted as a cure-all — is a particularly cruel hoax."

I said that would be quick.

Now back to the sun, the only truly safe source of power on the planet, the source of all life, and the engine of the earth.

Each of us can choose our energy future, and we just may have to as events force us to make a transition to a new energy paradigm. Richard Heinberg, mentioned above as the author of *The Party's Over*, has stated in his newest book, *Powerdown: Options and Actions for a Post Carbon World* (2004), that "the most likely trajectory for the energy transition will consist of the collapse of industrial civilization as we know it, probably occurring in stages over a period of several decades." That's not a very nice outlook!

He advocates becoming "preservationists" rather than "survivalists," meaning we can preserve a civilized way of life for ourselves and our communities by preparing now for a shared, not a selfish, "post carbon" lifestyle.

But whether we are sharing or selfish, the sun shines for all of us, and solarelectric technology is here and ready to go to work. Individually or collectively — the sun doesn't care — we can respond to the triple threats of global warming, energy security, and the end of oil. Or to quote Hermann Scheer once more, "The fight for renewables is a 'no' for fatalism and a 'yes' for an everlasting future and a spiritual hegemony." We can also live simply and use less; we don't *need* to personally send 10 to 20 tons of greenhouse gases into the atmosphere every year, as we in the West presently do. And "new energy for a new world" can be our guiding thought.

As long as the sun shines and photosynthesis grows our crops and the photovoltaic effect makes our power, what else do we need? When you've got solar power, every day is a sunny day. And best of all, solar power is *your* power.

# Glossary of Acronyms and Abbreviations

**ACORE:** American Council on Renewable Energy
**BP:** British Petroleum
**CIDA:** Canadian International Development Agency
**CRE:** Council for Renewable Energy
**DANIDA:** Danish International Development Agency
**DFCC:** Development Finance Corporation of Ceylon
**DOE:** Department of Energy
**EEAF:** Environmental Enterprises Assistance Fund
**ESD:** Energy Services Delivery (Project)
**ESMAP:** Energy Sector Management Assistance Program
**EVN:** Electricity of Vietnam
**GEF:** Global Environment Facility
**GNERI:** Gansu Natural Energy Research Institute
**GPV:** Gansu PV Company
**GTZ:** Gesellschaft für Technische Zusammenarbeit (German Technical Cooperation)
**IDA:** International Development Agency
**IFC:** International Finance Corporation
**IREDA:** Indian Renewable Energy Development Authority
**kV:** kilovolt
**kW:** kilowatt
**kWh:** kilowatt-hour
**LADWP:** Los Angeles Department of Water and Power

**mW:** megawatt
**NGO:** nongovernmental organization
**NORAID:** Norwegian Agency for International Development
**NOVIB:** Nederlandse Organisatie voor Internationale Bijstand (Oxfam Netherlands)
**NREL:** National Renewable Energy Laboratory
**NYPA:** New York Power Authority
**OPEC:** Organization of Petroleum Exporting Countries
**OPIC:** Overseas Private Investment Corporation
**PV:** photovoltaics
**PVMTI:** Photovoltaic Market Transformation Initiative
**RBF:** Rockefeller Brothers Fund
**REEEF:** Renewable Energy & Energy Efficiency Fund
**RESCO:** Renewable Energy Service Company
**SDC:** Solar Development Capital
**SDF:** Solar Development Foundation
**SDG:** Solar Development Group
**SEC:** Solar Electricity Company
**SEEDS:** Sarvodaya Economic Enterprises Development Services
**SELCO:** Solar Electric Light Company
**SERI:** Solar Energy Research Institute
**SHS:** solar home system
**SIDA:** Swedish International Development Agency
**SKDRDP:** Shree Kshetra Dharmastala Rural Development Program
**SSC:** solar service center
**UNDP:** United Nations Development Program
**UNEP:** United Nations Environment Program
**USAID:** United States Agency for International Development
**VBARD:** Vietnam Bank for Agriculture and Rural Development
**VWU:** Vietnam Women's Union
**Wp:** Watt Peak

# Bibliography

The following books are listed in order of importance to the issues of solar energy rather than alphabetically.

Scheer, Hermann. *The Solar Economy.* James & James, 1994.

Scheer, Hermann. *A Solar Manifesto.* Earthscan Publications, 2002.

Solar Energy International. *Photovoltaics Design and Installation Manual.* New Society Publishers, 2004.

Solar Living Center. *Solar Living Sourcebook.* New Society Publishers, 2005.

Schaeffer, John. *A Place in the Sun.* Chelsea Green, 1997.

Maycock, Paul D., and Edward N. A. Stirewalt, *Guide to the Photovoltaic Revolution.* Rodale Press, 1985.

Sklar, Scott, and Kenneth Sheinkopf. *Consumer Guide to Solar Energy.* Bonus Books, 2002.

Stapleton, Geoff, Lalith Gunaratne, and Peter JM Konings. *The Solar Entrepreneurs Handbook.* Global Sustainable Energy Solutions, 2002.

GTZ. *Basic Electrification for Rural Households.* Deutsche Gessellschaft für Technische Zusammenarbeit, GmbH (English), 1995.

Perlin, John. *From Space to Earth: The Story of Solar Electricity.* Harvard, 2002.

Butti, Ken, and John Perlin. *A Golden Thread: 2000 Years of Solar Architecture and Technology.* Van Nostrand Reinhold, 1980.

Berman, Daniel, and John O'Connor. *Who Owns the Sun?* Chelsea Green, 1996.

Heinberg, Richard. *The Party's Over.* New Society Publishers, 2003.

Heinberg, Richard. *Powerdown.* New Society Publishers, 2004.

Goodstein, David. *Out of Gas: The End of the Age of Oil.* W.W. Norton, 2004.

Roberts, Paul. *The End of Oil.* Houghton Miflin, 2004.

Gelbspan, Ross. *The Heat Is On.* Addison-Wesley Publishing Co., 1997.

Gelbspan, Ross. *Boiling Point.* Basic Books, 2004.

Oppenheimer, Michael, and Robert Boyle. *Dead Heat: The Race Against the Greenhouse Effect.* Basic Books, 1990.

Leeb, Stephen, and Donna Leeb. *The Oil Factor.* Warner Business Books, 2004.

Flavin, Christopher, and Nicholas Lenssen. *Power Surge: Guide to the Coming Energy Revolution.* W.W. Norton, 1994.

Schumacher, E.F. *Small Is Beautiful.* Harper & Row, 1973.

Ruppert, Michael C. *Crossing the Rubicon: The Decline of the American Empire at the End of the Age of Oil.* New Society Publishers, 2004.

Leggett, Jeremy. *The Carbon War: Dispatches from the End of the Oil Century.* Penguin, 1999.

Leggett, Jeremy, ed. *Climate Change and the Financial Sector: The Emerging Threat — The Solar Solution.* Gerling Akademie Verlag, 1996.

Schmidheiny, Stephan. *Financing Change: The Financial Community, Eco-efficiency, and Sustainable Development.* MIT Press, 1996.

Schmidheiny, Stephan. *Changing Course: A Global Business Perspective on Development and the Environment.* MIT Press, 1992.

Hanncock, Graham. *Lords of Poverty: The Power, Prestige, and Corruption of the International Aid Business.* Atlantic Monthly Press, 1989.

Rich, Bruce. *Mortgaging the Earth: The World Bank, Environmental Impoverishment, and the Crisis of Development.* Beacon Press, 1994.

Danaher, Kevin. *10 Reasons to Abolish the IMF & the World Bank.* Seven Stories Press, 2001.

Brown, Lester. *Eco-Economy: Building an Economy for the Earth.* W.W. Norton, 2001.

Brown, Lester. *Plan-B: Rescuing a Planet under Stress and a Civilization in Trouble.* W.W. Norton, 2003.

Hawken, Paul. *The Ecology of Commerce.* Harper Collins, 1993.

Hawken, Paul, Amory Lovins, and L. Hunter Lovins. *Natural Capitalism: Creating the Next Industrial Revolution.* Back Bay Books, 1999.

Korten, David C. *When Corporations Rule the World.* Barrett-Koehler, 1995.

Korten, David C. *The Post Corporate World: Life After Capitalism.* Barrett-Koehler, 1999.

Speth, Gustav. *Red Sky at Morning: American and the Crisis of the Global Environment.* Yale University Press, 2004.

de Soto, Hernando. *The Mystery of Capital: Why Capitalism Triumphs in the West and Fails Everywhere Else.* Basic Books, 2000.

Huffington, Arianna. *Pigs at the Trough.* Crown, 2003.

Perkins, John. *Confessions of an Economic Hit Man.* Barrett-Koehler, 2004.

Naisbit, John. *Global Paradox: The Bigger the World Economy, the More Powerful its Smallest Players.* Morrow, 1994.

Clarke, Arthur C. *How the World Was One: Beyond the Global Village.* Bantam, 1992.

Goldsmith, Edward, and Jerry Mander, ed. *The Case Against the Global Economy.* Sierra Club Books, 1996.

McLuhan, Marshall. *Understanding Media: The Extensions of Man.* McGraw Hill, 1964.

Greider, William. *One World Ready or Not: The Manic Logic of Global Capitalism.* Simon & Schuster, 1997

Phillips, Kevin. *Wealth and Democracy.* Broadway Books, 2002.

Bakke, Dennis. *Joy At Work: A Revolutionary Approach to Fun on the Job.* PVG, 2005.

Rifkin, Jeremy. *The Hydrogen Economy.* Tarcher-Penguin, 2003.

Reich, Charles A. *Opposing the System.* Crown, 1995.

Henderson, Hazel. *The Politics of the Solar Age.* Knowledge Systems, Inc., 1988.

Hendricks, Gay, and Kate Ludeman. *The Corporate Mystic: A Guidebook for Visionaries with Their Feet on the Ground.* Bantam, 1996.

Royal Institute of International Affairs; Wilkins, Gill. *Technology Transfer for Renewable Energy: Overcoming Barriers in Developing Countries.* Earthscan Publications, 2002.

Freeman, S. David. *Energy: The New Era.* Walker & Company, 1974.

Jonnes, Jill. *Empires of Light: Edison, Tesla, Westinghouse and the Race to Electrify the World.* Random House, 2003.

Hartman, Thom. *The Last Hours of Ancient Sunlight.* Three Rivers Press, 2000.

Kuntsler, James Howard. *The Long Emergency.* Atlantic Monthly Press, 2005.

Ewing, Rex. *Got Sun? Go Solar.* PixyJack Press, 2005

# Index

293